Interactive Mathematics Program®

IMP™

Integrated High School Mathematics

Small World, Isn't It?

Dan Fendel and Diane Resek
with
Lynne Alper and Sherry Fraser

Key Curriculum Press
Innovators in Mathematics Education

This material is based upon work supported by the National Science Foundation under award number ESI-9255262. Any opinions, findings, and conclusions or recommendations expressed in this publication are those of the authors and do not necessarily reflect the views of the National Science Foundation.

Key Curriculum Press
1150 65th Street
Emeryville, California 94608
www.keypress.com

10 9 8 7 6 5 4 3 2 11 10 09 08 07
ISBN-13 978-1-55953-921-X
ISBN-10 1-55953-921-6
Printed in the United States of America

Project Editor
Casey FitzSimons, Joan Lewis

Editorial Assistant
Jeff Gammon

Additional Editorial Development
Masha Albrecht, Mary Jo Cittadino

Art Developer
Ellen Silva

Production Editors
Caroline Ayres, Christine Osborne

Cover and Interior Design
Terry Lockman
Lumina Designworks

Production Management
Diana Jean Parks, Steve Rogers, Thomas Brierly

Art Editor
Laura Murray

Technical Graphics
Laurel Technical Services

Illustration
Taylor Bruce, Deborah Drummond, Tom Fowler, Evangelia Philippidis, Sara Swan, Diane Varner, Martha Weston, April Goodman Willy

Publisher
Steven Rasmussen

Vice President, Editorial and Production
Casey FitzSimons

Editorial Director
John Bergez

Textbook Product Manager
James Ryan

Mathematics Review
Rick Marks, Ph.D., Sonoma State University, Rohnert Park, California

Multicultural Reviews
Genevieve Lau, Ph.D., Skyline College, San Bruno, California
Arthur Ramirez, Ph.D., Sonoma State University, Rohnert Park, California
Marilyn Strutchens, Ph.D., University of Maryland, College Park, Maryland

Teacher Reviews
Daniel R. Bennett, Hoolehua, Hawaii
Maureen Burkhart, North Hollywood, California
Dwight Fuller, Shingle Springs, California
Daniel Johnson, San Jose, California
Brian Lawler, Aurora, Colorado
Brent McClain, Hillsboro, Oregon
Susan Miller, Philadelphia, Pennsylvania
Amy C. Roszak, Cottage Grove, Oregon
Carmen Rubino, Aurora, Colorado
Barbara Schallau, San Jose, California
Kathleen H. Spivack, New Haven, Connecticut
Wendy Tokumine, Honolulu, Hawaii

CONTENTS

Small World, Isn't It?

Appendix: Supplemental Problems 86

Small World, Isn't It?

DAYS 1–2
As the World Grows

DAYS 3–5
Average Growth

DAYS 6–10
All in a Row

DAYS 11–16
Beyond Linearity

DAYS 17–23
A Model for Population Growth

DAYS 24–29
The Best Base

DAYS 30–32
Back to the Data

DAYS 1–2

As the World Grows

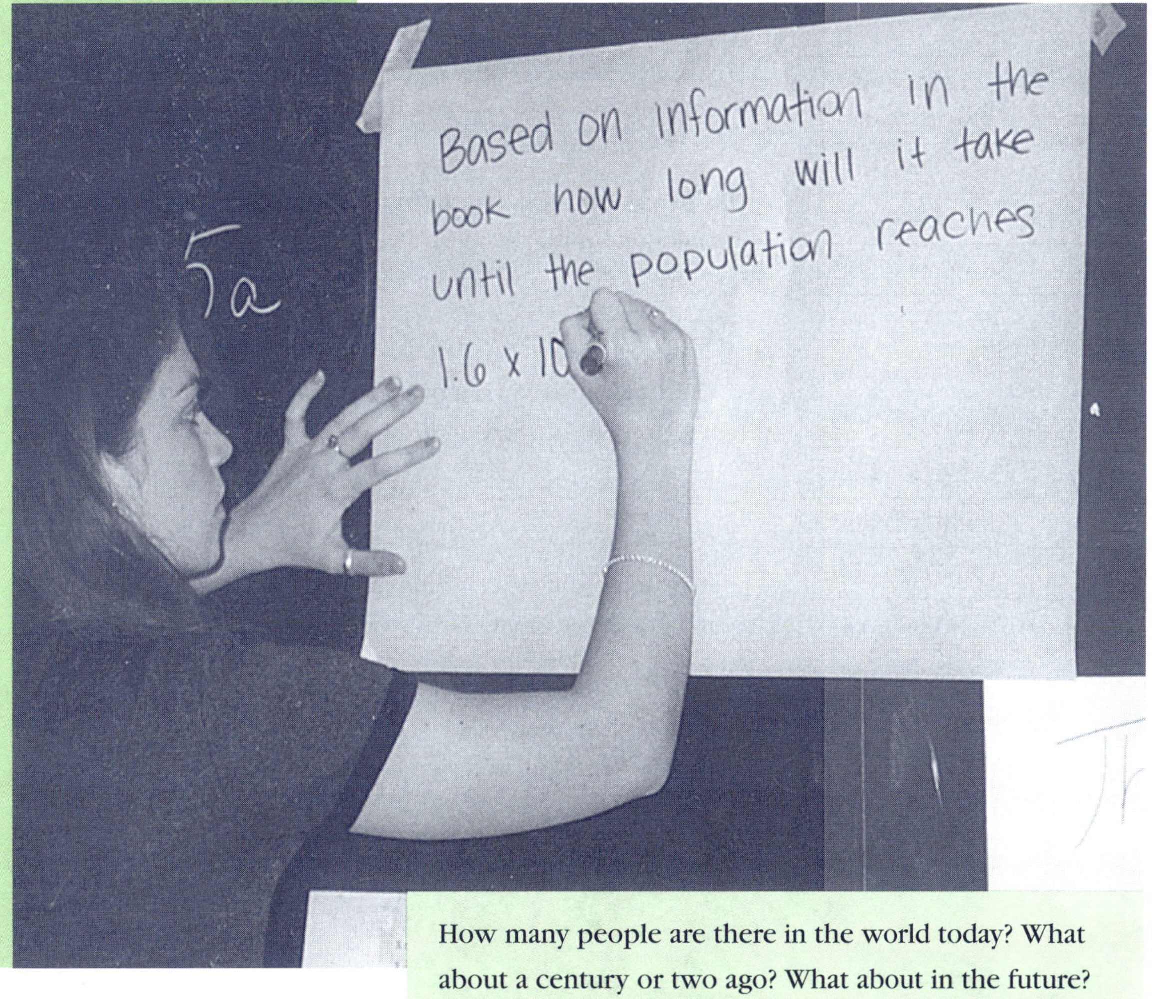

Lisa Newton rephrases the unit's central problem in her own words.

How many people are there in the world today? What about a century or two ago? What about in the future?

The central problem of this unit involves the analysis of world population trends. In the opening days of the unit, you will study population data since 1650, with the goal of determining how long it would take until people were "squashed up against one another."

CLASSWORK

Small World, Isn't It?

Year	Estimated Population
1650	470,000,000
1750	694,000,000
1850	1,091,000,000
1900	1,570,000,000
1950	2,510,000,000
1960	3,030,000,000
1970	3,680,000,000
1980	4,480,000,000
1985	4,870,000,000
1990	5,290,000,000
1995	5,730,000,000

Everyone knows that the world's population is increasing. The chart at the left shows the changes over the past several centuries.

If this pattern of data continued, how long do you think it would take until we were all squashed up against one another?

You may find these facts useful.

- The total surface area of the earth is approximately 196,930,000 square miles.
- Approximately 29.2% of the earth's total surface area is land.

As you consider this problem, make note of any difficulties you encounter or any issues you think need clarification.

Population and surface area data taken from *Information Please,* 1996 Almanac 49th edition Houghton-Mifflin p. 133.

POW 8

The More, the Merrier?

Isn't life more interesting when there are lots of people around? Then why should people be upset about the idea of a "population explosion"? Your task on this POW is to study some aspect of this rapid population growth and to find out why at least some people are concerned.

You should then write a report on your findings, which may include your own opinions and conjectures about why things are the way they are. Graphs, drawings, and tables are often helpful.

Here are some suggestions for topics. You may choose other topics for your report.

- Population growth and the food supply
- Pros and cons of limiting family size
- Comparison of population growth in different countries
- Population growth and pollution
- Ways to cope with the population explosion
- Traffic congestion

HOMEWORK 1

How Many of Us Can Fit?

In the activity *Small World, Isn't It?*, you are asked when we will all get "squashed up against one another." Of course, that probably won't ever happen, but you may be curious about how many people that would require.

1. How many square miles of land surface area are there on the earth? (Use the information provided in the activity *Small World, Isn't It?*)

2. How many square feet of land area are there on the earth? (Use the fact that 1 mile is equal to 5280 feet, so 1 square mile is 5280^2 square feet, which is 27,878,400 square feet.)

3. In 1995, the world population was about 5.73 billion people. If this population were spread evenly over the land area of the earth, about how many square feet of area would each person have, on average?

4. Suppose we interpret the phrase "squashed up against one another" to mean that each person has about one square foot to call his or her own. How many people would there need to be for us to really be squashed up against one another?

Data taken from *Information Please,* 1996 Almanac 49th edition Houghton-Mifflin p. 127.

How Many More People?

Part I: The Graph

One way to get some ideas about how the world population has grown over the centuries is through a graph.

1. Choose an appropriate scale and plot the data from *Small World, Isn't It?* Show all the specific data points given in that activity and label these data points on your graph with their coordinates.

Part II: Average Increases

2. a. Find the increase in population between the years 1650 and 1900.

   b. Explain how the increase you found in Question 2a shows up on the graph.

   c. Find the average increase in population *per year* between the years 1650 and 1900.

   d. What would the portion of the graph from 1650 to 1900 look like if the population increased by the same number of people each year of this time period?

3. a. Find the increase in population between the years 1900 and 1950.

   b. Explain how the increase you found in Question 3a shows up on the graph.

Continued on next page

c. Find the average increase in population *per year* between the years 1900 and 1950.

d. What would the portion of the graph from 1900 to 1950 look like if the population increased by the same number of people each year of this time period?

Part III: Making Comparisons

4. Which interval, 1650 to 1900 or 1900 to 1950, had the greater average increase in population per year? How could you answer this question simply by looking at your graph?

Growing Up

HOMEWORK 2

The accompanying graph shows the average height for boys ages 0 to 6 in the United States around the middle of the twentieth century.

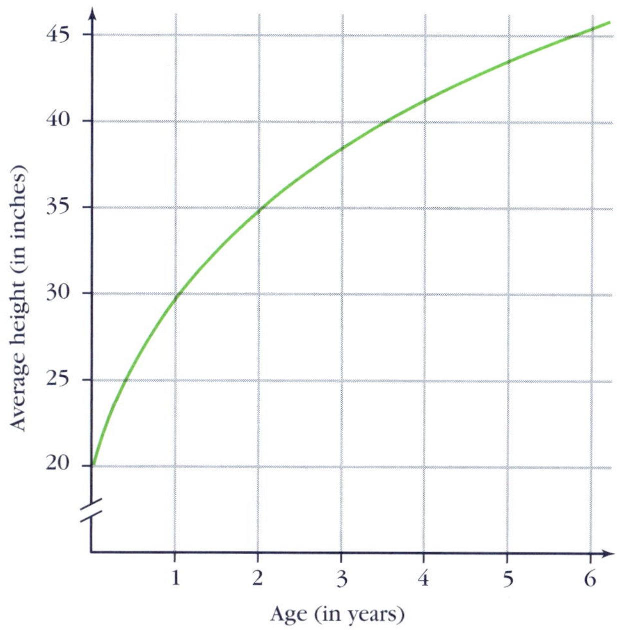

(Source: Department of Pediatrics, State University of Iowa, 1943.)

1. Suppose that the average amount of growth shown during the first year after birth were to continue during the second year. What would be the average height for boys on their second birthday?

2. The section of the graph for ages 3 to 6 is nearly straight. What does that mean in terms of the average growth for boys during those years?

3. **a.** How much does the average height for boys increase between ages 3 and 4?

b. Express your answer for Question 3a as a percentage of the average height for boys at age 3.

4. **a.** How much does the average height for boys increase between ages 5 and 6?

b. Express your answer for Question 4a as a percentage of the average height for boys at age 5.

c. Compare your answers for Questions 4a and 4b to your answers for Questions 3a and 3b.

DAYS 3–5

Average Growth

Leilani Juan and Liana DeGracia begin their investigation of growth and rates of change.

Most growth—of living things, of the economy, of organizations—is uneven, and population growth is no exception. Over the next few days, you'll look at some examples of growth and rates of change, focusing on how you can use graphs to help understand what's going on.

Story Sketches

HOMEWORK 3

1. This graph describes a wagon train's progress along the Overland Trail.

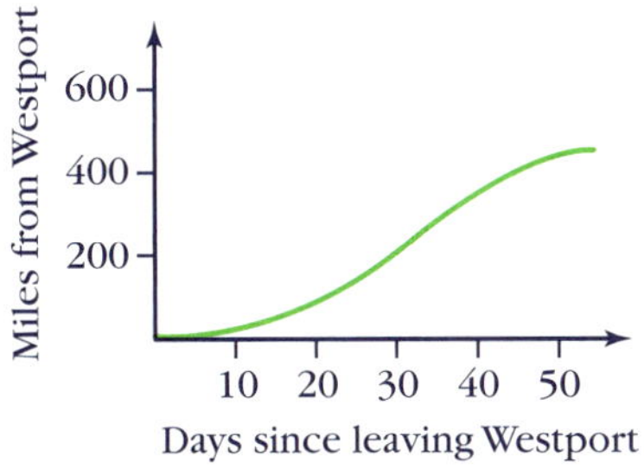

a. About how many miles per day did the wagon train go for the first 20 days? Explain how you got your answer.

b. What was the most rapid pace the wagon train achieved? Explain how you got your answer.

2. Tyler Dunkalot and his elementary school friends want to get basketball uniforms, and they are saving money so they can contribute to the purchase. One Friday afternoon, right after getting their weekly allowances, they each put some of the allowance into their piggy banks and count the savings so far.

Here's how much Tyler and his friends have that Friday (including the amount they just added) and how much each will add to savings every Friday afternoon from then on.

- Tyler has \$2 now and will add 50¢ each week.
- Robin has \$2 now and will add 70¢ each week.
- Mac has \$4 now and will add 70¢ each week.
- Dan has \$4 now and will add 30¢ each week.

Continued on next page

Draw graphs showing *accumulated savings* versus *time elapsed* for Tyler and each of his friends for a ten-week period. (Assume that they add their savings to their piggy banks each Friday and do not spend any of the saved money.)

Use the same set of axes for all four graphs, but use a different color to represent each youngster.

3. Compare your four graphs from Question 2. Describe how they are the same and how they are different.

4. For each of the youngsters, develop a formula or equation that shows how much money is in the piggy bank n weeks after the original count.

CLASSWORK

What a Mess!

Part I: The Mess

An oil tanker has broken up due to an explosion out at sea, and thousands of gallons of oil are spreading across the ocean. Linda Sue is flying overhead in an airplane and sees that the oil slick appears to be in the shape of a circle.

When Linda Sue first sees the oil slick, the radius of the circle is 70 meters. She flies overhead for a while and perceives that the radius is increasing at a rate of 6 meters per hour.

1. Make an In-Out table in which the *In* is the number of hours since Linda Sue first observed the oil slick and the *Out* is the radius of the oil slick after that many hours.

2. Draw a graph based on your In-Out table, and find a rule for your graph and table.

Part II: The Cleanup

Seven hours after Linda Sue first sees the oil spill with a radius of 70 meters, a cleanup operation begins. Rescue workers pump a special detergent into the center of the spill. As the detergent spreads, a circle of clean water develops inside the oil spill. The radius of this circle increases at 10 meters per hour.

3. Did the rescue team start soon enough and pump fast enough to eventually counteract the oil spill? If not, why not? If so, how many hours will it take until the spill is neutralized? (*Hint*: What was the radius of the oil spill when the cleanup operation began?)

HOMEWORK 4

Traveling Time

Abida has packed her bags and is ready to leave for college. She needs to catch an early train because she has a long way to travel.

Abida lives in Cincinnati, Ohio, and will be going to school in Philadelphia, Pennsylvania, 550 miles away. She is hoping to get to Philadelphia in time to see some historical landmarks. She figures that if she gets to Philadelphia by 3:00 p.m., she will have enough time to see the Liberty Bell and Independence Hall before they close for the evening.

1. If Abida catches a train leaving at 5:00 a.m., how many miles per hour must the train average for her to get there by 3:00 p.m.? (Cincinnati and Philadelphia are in the same time zone.)

2. If the 5:00 a.m. train averages 40 miles per hour for the first hour and a half, what speed must it average for the rest of the trip for Abida to make it to Philadelphia by 3:00 p.m.?

Continued on next page

3. For this question, suppose that the train actually averaged 50 miles per hour for the whole trip (which means that the trip took 11 hours altogether). That doesn't necessarily mean that the train traveled at a constant rate of 50 mph.

Make up a scenario in which the train averaged 50 mph for the trip but traveled at least two different speeds along the way. Be specific about speeds, times, and distances.

4. Abida's train left on time at 5:00 a.m. This graph shows one possibility for how far the train traveled as a function of the time elapsed.

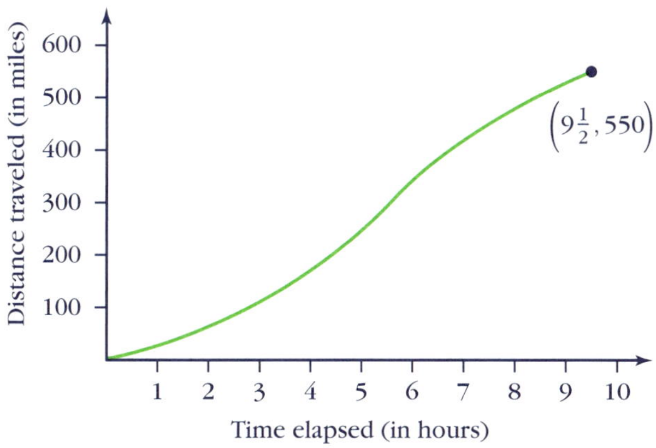

Use the graph to answer these questions.

a. Was the train's average speed faster during the first two hours of the trip or the last two hours?

b. During what one-hour period was the train's average speed the fastest?

c. If you had to pick the precise instant that the train was going the fastest, what would you choose and why?

CLASSWORK

Comparative Growth

Marvin was doing a study comparing the population growth of his hometown in the early twentieth century with its population growth in the early 1980s.

When he came across the two graphs shown here, Marvin noticed that the first graph appeared to be steeper than the second. He concluded that the population was growing faster at the beginning of the century than it was in the early 1980s.

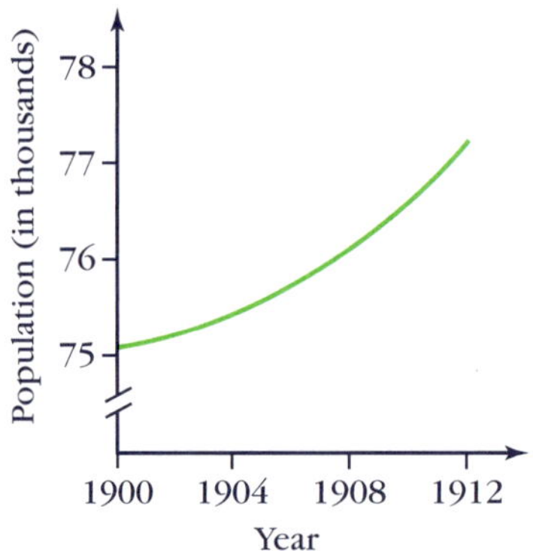

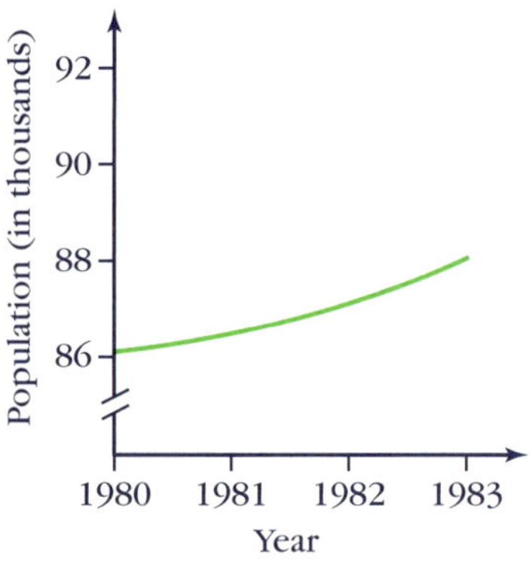

1. Use the first graph to calculate the average increase in population per year for the interval from 1900 to 1912.

2. Use the second graph to calculate the average increase in population per year for the interval from 1980 to 1983.

3. Do the results from Questions 1 and 2 support Marvin's conclusion? What does this suggest with regard to using graphs to compare rates of change?

If Looks Don't Matter, What Does?

In the activity *Comparative Growth*, you saw that you can't necessarily compare rates of growth in two graphs solely from the appearance of the graphs. In this assignment, you will look at some specific situations to decide what really matters.

1. The accompanying graph shows only two of the data points from the activity *Small World, Isn't It?*

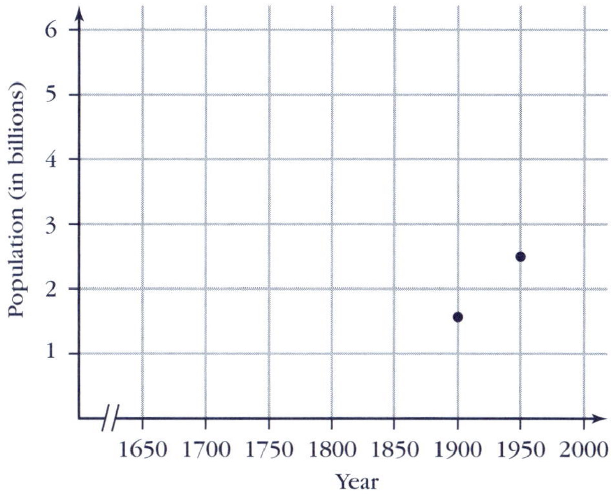

   a. Use the two given points to find the average population growth per year from 1900 to 1950, and explain the process you use.

   b. What would this portion of the graph look like if the population had grown by the same number of people each year?

Continued on next page

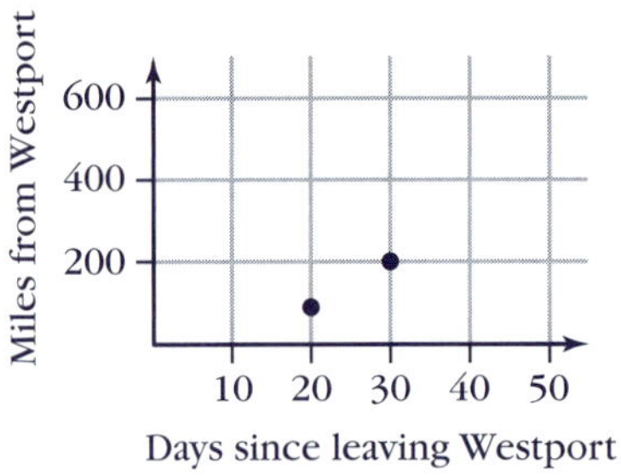

2. The graph at the left shows two points from the graph in Question 1 of *Homework 3: Story Sketches.* Use the graph to find the average distance traveled per day from Day 20 to Day 30, and explain the process you use.

3. Becky is a bit older than Tyler Dunkalot and his friends. The graph below shows the amount of money she had in her savings account at various times over the course of several weeks.

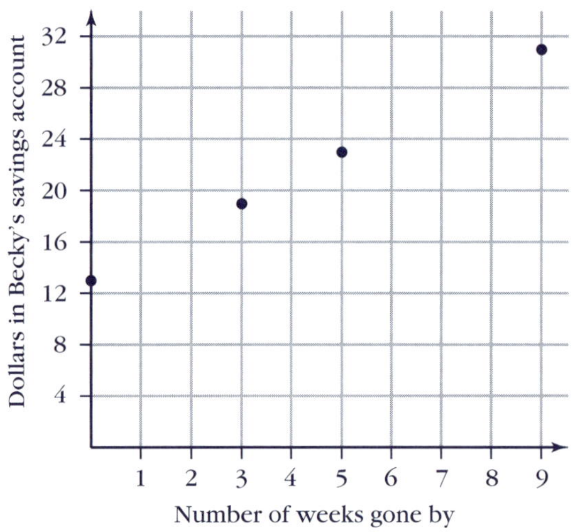

At what rate did the amount in Becky's savings account change per week? Explain how you got your answer from this graph.

Continued on next page

4. Find the rate of change of the function whose graph is shown here, and explain your work.

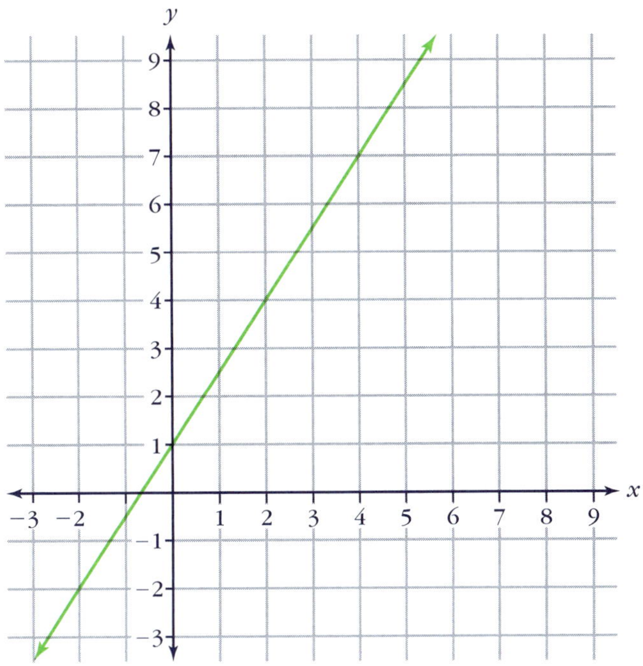

DAYS 6–10

All in a Row

Niall McNamara and Melyssa Brixner explore connections between graphs and formulas.

The simplest type of growth to study is constant growth, and that leads to work with linear functions. In the next segment of the unit, you will develop the important concept of *slope*, which is the primary tool for describing the rate of change for a linear function.

You'll use this concept to develop equations for straight lines, and you'll see whether a linear model for population growth can accurately predict future trends.

Formulating the Rate

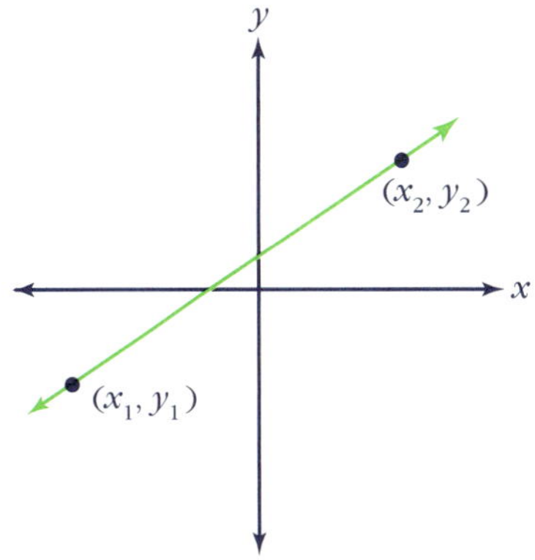

In each of the situations in *Homework 5: If Looks Don't Matter, What Does?,* you used coordinates of points on a graph to find a rate of change. In those problems, you were given specific numbers for the coordinates or could at least estimate the numerical values of the coordinates from the graph.

Now suppose that you have a function whose graph is a straight line and that (x_1, y_1) and (x_2, y_2) are two points on the graph. That is, instead of numerical values, you have variables representing each of the coordinates.

Find an expression for the rate of change of the function in terms of $x_1, y_1, x_2,$ and y_2.

HOMEWORK 6

Rates, Graphs, Slopes, and Formulas

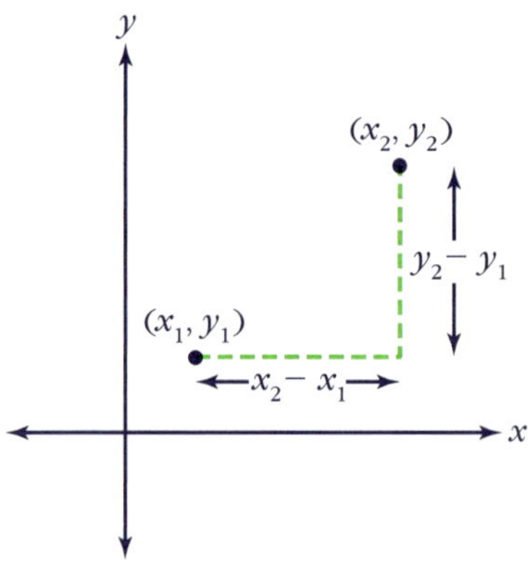

The concept of **slope** is an abstraction from the idea of a rate of change. Formally, the slope of the line connecting two points (x_1, y_1) and (x_2, y_2) is defined as the ratio

$$\frac{y_2 - y_1}{x_2 - x_1}$$

The numerator of this fraction is sometimes referred to as the **change in** $\boldsymbol{y}$ or the **rise.** The denominator is called the **change in** $\boldsymbol{x}$ or the **run.**

In these problems, you will work with both rates and slopes, exploring their connections to both graphs and formulas.

1. A jogger is moving along at a rate of 500 feet per minute. (That's about 6 miles per hour.)

   a. Use $t = 0$ as the time when the jogger starts, and plot a graph showing the distance the jogger has gone (in feet) as a function of t (in minutes).

   b. Find an equation for this function.

   c. Choose two points on your graph and use their coordinates to compute the slope of the graph.

2. A reservoir is partly filled and contains 200,000 cubic feet of water. Then water is added at a rate of 7000 cubic feet per hour.

   a. Use $t = 0$ as the time when the water begins to be added, and plot a graph showing the amount of water in the reservoir as a function of t. (You may want to "skip" part of the vertical axis as was done in the graph in *Homework 2: Growing Up.*)

Continued on next page

b. Find an equation for this function.

c. Choose two points on your graph and use their coordinates to compute the slope of the graph.

3. a. Plot the data in this In-Out table.

In	Out
3	18
5	26
10	46
16	70

b. Choose two points on your graph and use their coordinates to compute the slope of the line connecting them.

c. Find an equation to describe your graph in part a.

4. a. Choose two points on the line shown here and use their coordinates to compute the slope of the line.

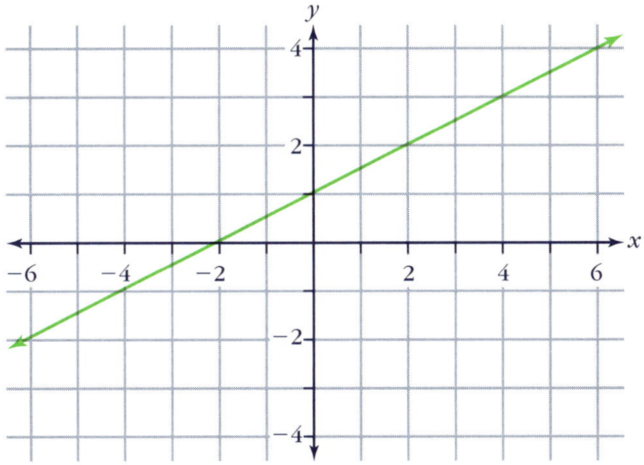

b. Find an equation for the line.

CLASSWORK

More About Tyler's Friends

In *Homework 3: Story Sketches,* you met Tyler Dunkalot and his friends, who were saving money to help buy basketball uniforms.

You may recall that two of Tyler's friends, Robin and Mac, both saved 70¢ per week, although at the start of their savings program, Robin had \$2 and Mac had \$4.

1. Graph both Robin's and Mac's savings as a function of the time elapsed. Treat these as *continuous* functions, as if Robin and Mac were each saving their allowance gradually throughout the week, at a constant rate. Use $t = 0$ to represent the moment when they started saving, and use the same set of axes for both graphs.

2. Find the slope of each of your graphs, and explain what you notice about the slopes.

3. Will the two graphs ever intersect? Explain your answer.

4. a. Find a formula for the amount of money each of the youngsters has at time t.

   b. How are these formulas related to your answers to Question 2?

Wake Up!

Getting going in the morning was probably hard for folks on the Overland Trail (just as it may be for you today), so coffee was a highly valued commodity.

The Cazneau family started out with a supply of 30 pounds of coffee. When they arrived at Sutter's Fort in California, after a 188-day journey, they had only 3 pounds left.

1. Draw a graph showing the amount of coffee remaining as a function of the amount of time elapsed since the start of the journey. For the sake of simplicity, assume that the Cazneaus consumed coffee at a constant rate.
2. Find the average amount of coffee consumed per day.
3. Develop an equation for the function represented by your graph in Question 1.
4. Find the slope of your graph from Question 1.
5. Suppose that at the beginning of their journey, the Cazneaus were 1600 miles from Sutter's Fort. As stated previously, their journey took a total of 188 days.
    a. How many miles per day did they travel? (For simplicity, assume that they traveled the same distance each day.)
    b. Of course, as the Cazneaus traveled, their distance from Sutter's Fort decreased. Write a formula expressing the distance remaining as a function of the number of days they'd been traveling.
    c. Make a graph of your function from Question 5b.
    d. Find the slope of your graph from Question 5c.

HOMEWORK 8

California, Here I Come!

According to the U.S. census, the population of California in 1850 was about 92,600. During the 1850s, with the great westward migration, the population grew substantially. In 1860, the census population was 380,000.

1. What was the average annual population increase during the 1850s (that is, from 1850 to 1860)?

2. If California's population had continued to increase after 1860 by the same annual amount as it averaged during the 1850s, what would the population have been in these years?

   a. 1900

   b. 1950

   c. 1990

3. Generalize your results from Question 2. That is, develop a formula for the population of California in the year X (based on the assumption that the population of California had continued to increase after 1860 by the same annual amount as it averaged during the 1850s).

4. According to the official U.S. census, California's population in 1990 was 29,760,021. How does this amount compare to the amount you found in Question 2c? What do you think accounts for the difference between your prediction for 1990 and the actual population?

Historical note: At the time of the Gold Rush (the 1850s), the U.S. Constitution stated that taxes and representation

Continued on next page

in Congress would be apportioned among the states "according to their respective numbers, which shall be determined by adding to the whole number of free persons, including those bound to service for a term of years, and excluding Indians not taxed, three fifths of all other persons." Thus, Native Americans were not counted at all unless they were taxed, and slaves were counted as three-fifths of a person each. The Fourteenth Amendment (1868) changed this to count everyone except "Indians not taxed."

There is still controversy today about the accuracy of the census, with some people claiming that inner-city or rural populations are undercounted.

POW 9

Planning the Platforms

The Platform Display

River City is getting ready for the big Fourth of July band concert that precedes the fireworks. The concert is always a major event, but this year the band leader, Kevin, plans to make it bigger and better than ever.

Kevin wants to have each of the baton twirlers standing on an individual platform, as shown here.

The baton twirlers will toss batons up and down to one another. Kevin wants the difference in height from one platform to the next to be the same in each case.

Kevin's Decisions

Kevin has several decisions to make.

- He needs to decide on the number of platforms. (Kevin isn't sure how many of his baton twirlers will be good enough to perform by the Fourth of July.)
- He needs to decide on the height of the first platform. (This will depend on how tall the baton twirler on the first platform is, and Kevin hasn't decided who the first baton twirler will be.)
- He needs to decide on the difference in height from one platform to the next. (Kevin doesn't know yet how high the twirlers will be able to toss their batons.)

Camilla's Dilemma

Camilla is in charge of building and decorating the structure. She needs a permit from the city to build the

Continued on next page

structure, so she needs to know how high the tallest platform will be.

She also plans to hang a colorful strip of material on the front of each platform. Each strip will reach from the top of the platform to the ground. (The width of the material is the same as the width of each platform, so she needs only one strip per platform.) She needs to know the total length of material that she should buy.

Camilla is going crazy, because she can't do her job until Kevin makes his decisions.

Your Task

You are Camilla's assistant, and she has asked you to be ready to give her the information she needs as soon as Kevin makes up his mind.

Your task in this POW is to create two formulas that will allow you to do this instantly. One formula should tell you the height of the tallest platform. The other should tell you the total length of material that Camilla will need. Your formulas should give these results in terms of the number of platforms, the height of the first platform, and the difference in height between adjacent platforms.

Write-up

1. *Problem Statement:* State the mathematical essence of the problem, independent of the context of the platforms, batons twirlers, and so on.
2. *Process:* Give details of any specific examples you worked out, and explain how those examples helped you develop the formulas.
3. *Solution:* Give the two formulas and explain why they work.
4. *Evaluation*
5. *Self-assessment*

HOMEWORK 9

Points, Slopes, and Equations

In *Homework 8: California, Here I Come!,* you saw that knowing two data points allows you to develop a linear model for population growth. This makes geometric sense, because two points determine a straight line.

When you actually wrote the equation, you may have worked primarily with the slope (the rate of growth) and one of those points. This also makes geometric sense, because a line can also be determined by knowing one point and the "steepness" of the line (the slope).

The problems in this assignment are similar to your work in that homework, except that they do not involve population. For each problem in Part I, you are given the slope along with one point. For each problem in Part II, you are given two points. For Part II, you may want to find the slope as a first step in order to write the equation. Part III examines the special cases of horizontal and vertical lines.

Part I: Point-Slope Equations

1. Find an equation for the line with slope 5 that goes through the point $(3, 2)$.

2. Find an equation for the line with slope 10 that goes through the point $(-4, 7)$.

3. Find an equation for the line with slope -3 that goes through the point $(5, -4)$.

4. Find an equation for the line with slope $\frac{2}{3}$ that goes through the point $(4, 6)$.

Continued on next page

Part II: Two-Point Equations

5. Find an equation for the line that goes through the points $(6, 2)$ and $(8, 8)$.

6. Find an equation for the line that goes through the points $(1, 5)$ and $(3, -1)$.

7. Find an equation for the line that goes through the points $(-7, 2)$ and $(-2, 5)$.

Part III: Horizontal and Vertical Lines

8. Find an equation for the horizontal line that goes through the point $(5, 1)$. What is the slope of this line?

9. Find an equation for the vertical line that goes through the point $(2, -6)$. What is the slope of this line?

CLASSWORK

The Why of the Line

You've seen that any pair of points on a given line will lead to the same slope. But why is this so?

Consider this diagram, which shows four points—A, B, C, and D—on the same line. The diagram also shows right triangles using the segments $\overline{AB}$ and $\overline{CD}$ as their hypotenuses. The lengths of the legs of these right triangles are represented using the variables r, s, t, and u, as shown.

Use ideas from geometry to explain why the slope for the pair of points A and B must be the same as the slope for the pair of points C and D. That is, explain why the ratio $\frac{r}{s}$ must be the same as the ratio $\frac{t}{u}$.

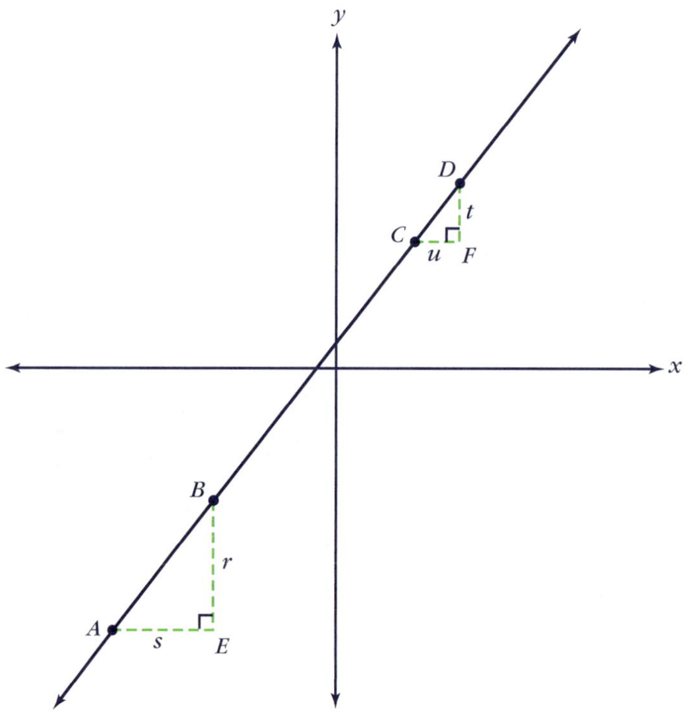

HOMEWORK 10

Return of the Rescue

You may recall this situation from the activity *To the Rescue* in the Year 2 unit *Solve It!*

> A helicopter is flying to drop a supply bundle to a group of firefighters who are behind the fire lines. At the moment when the helicopter crew makes the drop, the helicopter is hovering 400 feet above the ground.

The principles of physics that describe the behavior of falling objects state that when an object is falling freely, it goes faster and faster as it falls. In fact, these principles provide a specific formula describing the object's fall, which can be expressed this way.

> If an object is dropped from a height of N feet, then $h(t)$, its height off the ground (in feet) t seconds after being dropped, is given by the equation $h(t) = N - 16t^2$.

So in the case of the falling supplies, the formula is $h(t) = 400 - 16t^2$, because the supply bundle is 400 feet off the ground when it starts to fall.

1. What is $h(3)$? That is, what is the height of the supply bundle three seconds after it is dropped?
2. How far has the supply bundle fallen during those three seconds?
3. What is the average speed of the supply bundle during those three seconds?
4. How long does it take for the supply bundle to reach the ground?

DAYS 11–16

Beyond Linearity

Lars Lundgren, Kim Watt, and Betsy Hooper prepare to present what they mean by "the speed at the exact moment."

Slope is an excellent concept for working with straight-line graphs. But what about rates of change that are not constant?

With the activity *The Instant of Impact,* you will begin an exploration of the meaning of *instantaneous rate of change.* Over the next week or so, you will refine this idea and explore it from several perspectives.

CLASSWORK

The Instant of Impact

As described in *Homework 10: Return of the Rescue,* a supply bundle has been dropped from a helicopter. The bundle's height off the ground t seconds after being dropped is given by the equation

$$h(t) = 400 - 16t^2$$

with $h(t)$ measured in feet.

Suppose that the bundle can withstand the impact of hitting the ground only for speeds up to 165 feet per second. To determine whether the bundle will survive the fall, you need to find out how fast it is traveling when it hits the ground.

Keep careful track of your computations on these questions, so you can compare results from question to question.

1. a. How far does the bundle fall during its last two seconds (that is, during the time from $t = 3$ to $t = 5$)?

   b. What is the bundle's average speed during that two-second interval?

2. a. How far does the bundle fall during the last second before it hits the ground?

   b. What is the bundle's average speed during this last second?

3. What is the bundle's average speed during the last half-second before it hits the ground?

4. What is the bundle's average speed during the last tenth of a second before it hits the ground?

5. What is the bundle's speed at the exact moment it hits the ground?

HOMEWORK 11

Doctor's Orders

Larry is a stunt diver. His most famous dive is off a cliff into the ocean near the town of Poco Loco.

Unfortunately, Larry injured his wrist recently, and the impact of hitting the water at high speeds could do further damage. His doctor recommends that he do no dives in which the speed of his entry into the water is greater than 60 feet per second.

The cliff at Poco Loco is 62 feet high, but Larry always begins his dive with a jump, so he actually starts his fall from a height of 64 feet. Therefore, his height above the ocean is given by the formula

$$h(t) = 64 - 16t^2$$

where t is the time, in seconds, from when he begins his fall, and $h(t)$ is his height, in feet, above the ocean.

1. How high above the ocean is Larry one second after he begins his fall?
2. What is the value of t when Larry hits the water?
3. What is Larry's average speed during the final second of his dive?
4. What is Larry's average speed during the final half-second of his dive?
5. Can Larry perform his lucky dive without violating his doctor's instructions? Explain.

Photo Finish

You may recognize this situation from the activity *Where's Speedy?* in the Year 2 unit *Solve It!*

> Speedy is the star runner for her country's track team. Among other things, she runs the last 400 meters of the 1600-meter relay race.
>
> A sports analyst studied the film of a race in which Speedy competed. The analyst came up with this formula to describe the distance Speedy had run at a given time in the race:
>
> $$m(t) = 0.1t^2 + 3t$$

In this formula, $m(t)$ gives the number of meters Speedy had run after t seconds of the race, with both time and distance measured from the beginning of her 400-meter segment of the race. (This formula might not be very accurate, but on this activity you should treat it as if it were completely correct.)

1. How long did it take Speedy to finish the race? That is, how long did it take her to run 400 meters? Explain how you found your answer.

2. A photograph was taken of Speedy at the instant she crossed the finish line. The photo shows a slight blur, so you know Speedy was going pretty fast, but you can't tell the speed at which she was going at the instant the photo was taken.

   Find Speedy's speed at that instant.

3. **a.** Find Speedy's speed at three other instants during the race.

   b. Was there an instant when Speedy was going exactly 10 meters per second? If so, when was that instant?

HOMEWORK 12

Speed and Slope

Part I: An Instantaneous Summary

You know that *average speed* can be defined by the simple formula

$$\text{average speed} = \frac{\text{distance traveled}}{\text{time elapsed}}$$

but you've seen that *instantaneous speed* is a more complicated idea.

Summarize what you have learned about how to calculate instantaneous speed from *The Instant of Impact,* from *Homework 11: Doctor's Orders,* and from *Photo Finish.*

Part II: A Linear Review

In recent activities, you have looked at how rate of change might be calculated for nonlinear situations. But you shouldn't forget how to use constant rates and the slope of a straight line, whether in the context of a real-life situation or in terms of a graph. Here is one problem of each type.

1. Ruth keeps track of rainfall in her area, by watching the level of water in a barrel outside her house. A steady rainstorm started one day, and at one point, she observed that the water level was at 420 millimeters. Three hours later, the level had increased to 438 millimeters.

   Assume that the rain continued steadily, so that the water level rose at the same rate throughout that three-hour period. Use t to represent the time elapsed (in hours) from Ruth's first measurement, and write a function expressing the water level in terms of t.

2. Find an equation for the line that goes through the points $(5, -8)$ and $(13, 4)$.

ZOOOOOOOOM

Consider the function $m(t) = 0.1t^2 + 3t$, which described the distance Speedy had run in terms of time elapsed.

Graph this function on the graphing calculator, adjusting the viewing window so that your graph includes the point (50, 400). (Recall that she completed her 400-meter run in 50 seconds, so this point is on the graph.)

Then zoom in on the point (50, 400) until the graph on your screen looks pretty much like a straight line.

1. a. Use the calculator's trace feature to find the coordinates of two points on this apparent straight line, and write down those two coordinate pairs.

   b. Find the slope of the line connecting those points.

2. What does the slope you found in Question 1b mean in terms of Speedy's race?

HOMEWORK 13

The Growth of the Oil Slick

The idea of an instantaneous rate of change doesn't apply only to speed. The questions here concern the rate of growth of the area of the oil slick (from *What a Mess!*).

You may recall that when Linda Sue first saw the circular oil slick, the radius of the circle was 70 meters. She also noticed that the radius was increasing by 6 meters per hour. This means that the radius can be described by the formula $r = 70 + 6t$ (where t is the number of hours since Linda Sue first saw the oil slick). Therefore, the area in square meters of the oil slick after t hours is given by the function

$$A(t) = \pi(70 + 6t)^2$$

1. What was the area covered by the oil slick when Linda Sue first saw it (at $t = 0$)?

In the remaining questions, express the rate of growth of the oil slick in square meters per hour.

2. a. What was the average rate at which the oil slick grew during the first two hours after Linda Sue's initial observation?

   b. What was the average rate at which the oil slick grew during the first half-hour after Linda Sue's initial observation?

   c. What was the average rate at which the oil slick grew during the 15 minutes *before* Linda Sue's initial observation? (*Suggestion:* Think of 15 minutes as one-fourth of an hour.)

3. At what rate was the oil slick growing at the instant when Linda Sue first saw it?

Speeds, Rates, and Derivatives

The derivative of a function at a point is one of the basic concepts of calculus. If a function is defined by the equation $y = f(x)$ and (a, b) is a point on the graph, then the **derivative of f at (a, b)** can be thought of in at least two ways.

- It is the slope of the line that is tangent to the curve at (a, b).
- It is the instantaneous rate at which the y-value of the function is changing as the x-value increases through $x = a$.

Note: We often refer to this as "the derivative at $x = a$" rather than as the derivative at (a, b).

This assignment gives you a chance to work with this new idea in connection with some familiar situations. Keep in mind that you can find derivatives by using smaller and smaller intervals around a particular value of a.

Continued on next page

1. The function $h(t) = 400 - 16t^2$ gives the height of the supply bundle (in feet) t seconds after it is dropped.

   a. Find the derivative of this function at the point $(3, 256)$.

   b. What does the answer to Question 1a tell you about the speed at which the supply bundle is falling?

2. The function $A(t) = \pi(70 + 6t)^2$ gives the area of the oil slick (in square meters) t hours after Linda Sue first saw it.

   a. Find the rate at which the area was growing exactly one hour after Linda Sue first saw it.

   b. Express the answer to Question 2a as a derivative.

3. *Homework 7: Wake Up!* described the Cazneau family's consumption of coffee. The function $f(d) = 30 - 0.14d$ gives a good approximation of the amount of coffee (in pounds) the Cazneau family had left after d days.

   For this assignment, you should treat this as a continuous function, rather than as a discrete function. That is, assume that d need not be a whole number.

   a. Because $f(5) = 29.3$, the graph of this function goes through the point $(5, 29.3)$. What is the derivative of the function at this point?

   b. What does the answer to Question 3a tell you about the rate at which the Cazneaus drank coffee?

   c. Pick a point on the graph of the function other than $(5, 29.3)$, and find the derivative at that point.

Zooming Free-for-All

You saw in *ZOOOOOOOOM* that if you zoom in on the calculator graph of the function $m(t) = 0.1t^2 + 3t$, the graph on the screen quickly begins to look like a straight line. (Of course, you have to take into account that on a calculator screen, even the graph of a linear function might not look perfectly straight.)

Your task in this activity is to investigate whether this phenomenon occurs for other functions as well.

1. Start with the calculator graph of the function $h(t) = 400 - 16t^2$, representing the height of the falling supply bundle. Choose a point on this graph and zoom in on that point.
   a. Does the graph begin to appear straight?
   b. If the graph does appear straight, what does the slope of that apparent straight line mean in terms of the falling supply bundle?
2. Experiment with other functions and other points. Pick a function, choose a point on its calculator graph, and zoom in on that point. (Don't worry about whether you can find a meaningful real-life situation for the function.)
   a. Does the graph begin to appear straight?
   b. If the graph does appear straight, what does a straight line with that slope through your given point represent in terms of the graph?
3. Can you think of a graph and a point on the graph so that no matter how much you might zoom in on that point, the graph would still not appear straight? You might try to sketch such a graph by hand, even if you can't find a formula for it.

HOMEWORK 15

On a Tangent

A *secant* for the graph of a function is the line (or line segment) connecting two points on the graph. A *tangent* is a line that "just touches" the graph at a point. This assignment explores the two concepts and their connections with derivatives.

1. Consider the function f defined by the equation $f(x) = 0.5x^2$.

   a. Sketch the graph of this function, with the scale on your x-axis going from -1 to 3. Use a full-size sheet of graph paper for this sketch so that you will be able to get enough detail in Question 1c.

   b. Label the point $(2, 2)$ on your sketch.

   c. The points listed here are also on the graph. In each case, draw the secant connecting the given point to $(2, 2)$ and find the slope of that secant.

      i. $(0, 0)$

      ii. $(1, 0.5)$

      iii. $(1.5, 1.125)$

      iv. $(1.9, 1.805)$

   d. Draw the line that is tangent to the graph at $(2, 2)$. Then estimate the slope of that line and explain your reasoning.

   e. Find the derivative of the function f at the point $(2, 2)$.

Continued on next page

2. The accompanying diagram shows the graph of a function.

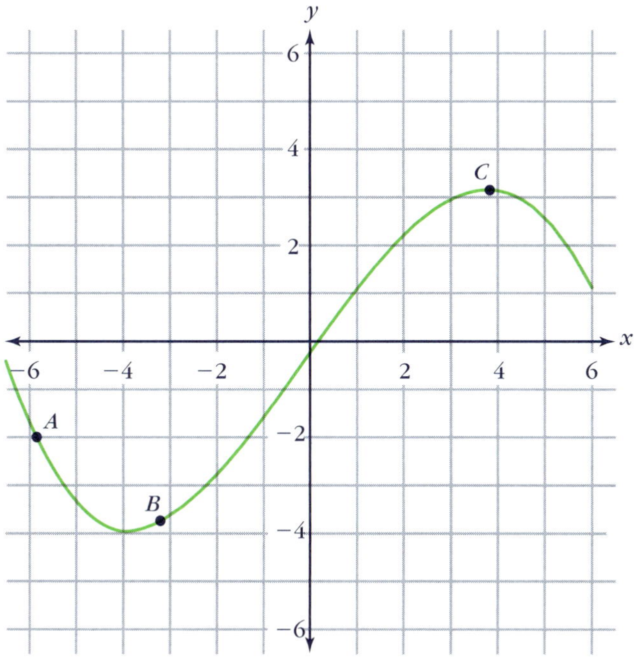

a. Make a copy of this graph.

b. Draw the tangent lines to the graph at each of the points *A, B,* and *C*.

c. Use your work from Question 2b to estimate the derivative of the function at each of the points *A, B,* and *C*.

POW 10

Around King Arthur's Table

King Arthur was the ruler in Camelot who had all of those knights and a round table. He loved inviting the knights over for parties around his round table.

If there was something pleasant that the king could give to only one knight (an extra dessert, a dragon to chase, and so on), he had them play a game to determine who would get it.

The game went like this.

First, King Arthur put numbers on the chairs beginning with 1 and continuing around the table, with one chair for each knight. He had the knights sit down so that every chair was occupied. King Arthur himself remained standing.

Then King Arthur stood behind the knight in chair 1 and said, "You're In." Next, he moved to the knight in chair 2 and said, "You're Out," and that knight left his seat and went off to stand at the side of the room to watch the rest of the game. Next, he moved to the knight in chair 3 and said, "You're In." Then he said, "You're Out" to the knight in chair 4, and that knight left his seat and went to the side of the room.

The king continued around the table in this manner. When he came back around to the knight in chair 1, he said either "You're In" or "You're Out," depending on what he had said to the previous knight. (If the previous knight was "In," then the knight in chair 1 was now "Out," and vice versa.)

Continued on next page

The king kept moving around and around the table, alternately saying, "You're In" or "You're Out" to the knights who remained at the table. (If a chair was now empty, he simply skipped it.) He continued until only one knight was left sitting at the table. That knight was the winner.

Your Task

Of course, the number of knights varied from day to day, depending on who was sick, who was out chasing dragons, and so on. Sometimes there were only a handful, and sometimes there were over a hundred!

Here's the big question of the POW.

> *If you knew how many knights were going to be at the table, how could you quickly determine which chair to sit in so that you would win?*

Your task is to develop a general rule, formula, or procedure that will predict the winning seat in terms of the number of knights present. Be sure to explain why your rule works.

Write-up

1. *Problem Statement*
2. *Process*
3. *Solution*
4. *Evaluation*
5. *Self-assessment*

HOMEWORK 16

What's It All About?

The idea of "the derivative of a function at a point" is a very important one in mathematics. In fact, this idea plays a key role in calculus.

You will be using this idea more in the unit. Thus, it would be a good idea for you to pull together what you know about it now, so that you will be ready to build on that knowledge.

Write down what you have learned about the idea of a derivative so far. Focus on these issues.

- What it means
- How you calculate it
- How it relates to other important ideas

Be sure to include specific examples.

DAYS
17–23

A Model for Population Growth

Erika Cohen and Alex Ryane are writing equations based upon patterns they observed from the tables they completed for "Slippery Slopes."

You've seen that linear functions do not necessarily provide a good model for population growth. So what kind of function should you use?

In this section of the unit, you'll discover that for a certain family of functions, the derivative has a special property that makes it an excellent candidate to use in modeling population growth.

How Much for Broken Eggs?!!?

Do you remember a Problem of the Week called *The Broken Eggs*? (It was the first POW in Year 1.) As you may recall, it involved a farmer whose cart was hit as she was taking her eggs to market. Though she wasn't injured, her eggs got broken, and you spent a while figuring out how many eggs there were.

The cost of replacing 301 eggs may not have amounted to much then, but as prices rise, it could involve a lot of money. In this activity, you'll investigate what would happen if prices went up by the same percentage each year.

In this activity, you should assume that at the end of 1990, a dozen eggs cost 89¢ and that prices went up 5% every year. (In a situation like this, the figure of 5% is called the *rate of inflation.*)

Continued on next page

The first two questions are intended to get you started in your analysis of the situation.

1. a. How much did a dozen eggs cost at the end of 1991?

   b. How much did the price go up during 1991?

2. a. How much did a dozen eggs cost at the end of 1992?

   b. How much did the price go up during 1992?

Now gather similar information for other years, until you think you understand what's happening. Then answer these questions, assuming that the 5% inflation rate continues.

3. How much will a dozen eggs cost at the end of the year 2100? Explain your answer.

4. In what year will a dozen eggs first cost over $100?

HOMEWORK 17

Small but Plentiful

How Much for Broken Eggs?!!? involves growth in prices, based on a situation from a Year 1 POW. This assignment involves a type of population growth and is somewhat like *POW 1: Growth of Rat Populations,* in the unit *Fireworks.*

As you might recall, that POW was pretty complicated. You had to keep track of males and females, of different generations of rats, of when the females were ready to give birth, and so on. The situation in this assignment is much simpler.

Imagine a microscopic creature like an amoeba. Suppose that whenever one of these creatures gets to a certain size, it splits into two. Then these two each grow, and

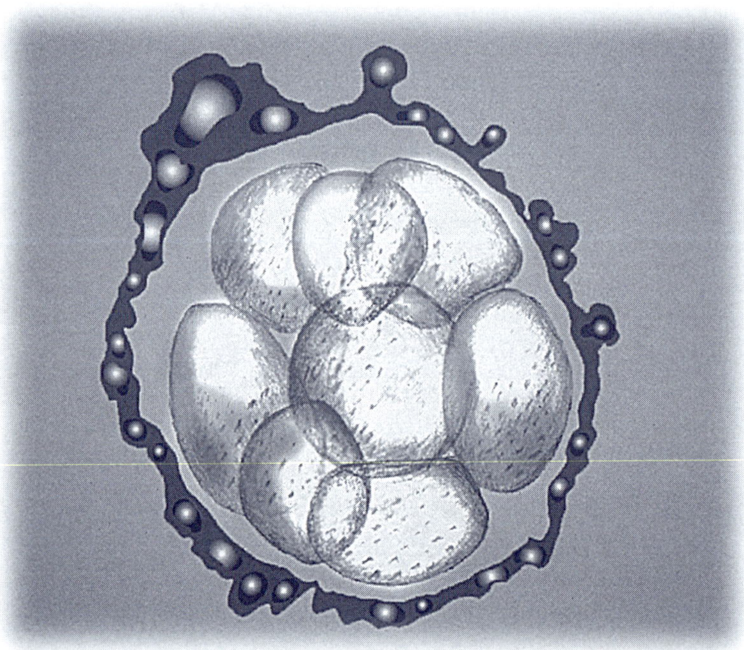

Continued on next page

when they get big enough, they each split into two, making four altogether. And so on.

Suppose further that at 12:01 a.m. on January 1 (just after midnight), there is one such tiny creature. And finally, suppose that it takes exactly 12 hours for such a creature to get big enough to split into two, with the first split taking place at 12:01 p.m. (just after noon) on January 1.

Your task is to find a general rule for figuring out the number of creatures at a given time. (*Note:* In this model, these creatures never die—they simply split into two.)

1. Begin with specific examples by figuring out how many creatures there are at each of these times.
   a. At 12:01 p.m. on January 1
   b. At 12:01 a.m. on January 2
   c. At 12:01 a.m. on January 5
   d. At 12:01 a.m. on January 31
2. Now find a general formula that tells how many creatures there are at 12:01 a.m., d days after the start of the experiment. (At 12:01 a.m. on January 1, $d = 0$.)
3. What will be the first day when there are over 1 million creatures?

HOMEWORK 18

The Return of Alice

In the Year 2 unit *All About Alice,* Lewis Carroll's fictional Alice could change her height by eating various kinds of special cake.

Each ounce she ate of a given type of cake multiplied her height by a particular factor, and Alice named the different kinds of cake to match their effect. For example, if the cake multiplied her height by a factor of 3, she called it "base 3 cake."

1. What would Alice's height be multiplied by if she ate each of these amounts of the particular types of cake? Write your answers using exponential expressions.

a. 4 ounces of base 3 cake

b. 5 ounces of base 2 cake

c. x ounces of base 7 cake

2. Suppose Alice ate 4 ounces of base 2 cake and then 3 more ounces of the same type of cake.

a. Use this situation to explain the equation $2^4 \cdot 2^3 = 2^7$.

b. Explain the equation $2^4 \cdot 2^3 = 2^7$ in terms of repeated multiplication.

3. Explain the equation $(4^6)^7 = 4^{6 \cdot 7}$ in each of these ways.

a. In a situation involving Alice

b. In terms of repeated multiplication

Alice realized that she could express "cake questions" using exponential equations. For example, if she wanted to know how much base 2 cake to eat to multiply her height by 32, she would ask herself, "What's the solution

Continued on next page

to the equation $2^x = 32$?" She also realized that the solution to this equation could be expressed as a logarithm, namely, $\log_2 32$.

4. Find the value of $\log_2 32$ by solving the equation $2^x = 32$.

5. For each of these questions, write an exponential equation to represent the situation, find the numerical solution to the equation, and write the solution as a logarithm.

a. How much base 3 cake should Alice eat to multiply her height by 81?

b. How much base 2 cake should Alice eat to multiply her height by 128?

c. How much base 5 cake should Alice eat to multiply her height by 93? (For this example, give your numerical solution to the nearest tenth.)

CLASSWORK

Slippery Slopes

The derivatives of exponential functions show a very interesting pattern. Your task in this activity is to find the pattern.

1. Start with the exponential function defined by the equation $y = 2^x$.

a. Your first task is to create an In-Out table like the one shown here.

x-value	y-value	Derivative

Use these two steps to get a row of the table.

- Pick a whole-number value for x and find the y-value that goes with it.
- Get a good approximation for the derivative of the function at the point on the graph of the function $y = 2^x$ that is represented by your x- and y-values.

Go through these two steps for several points to get several rows for the table.

b. Once you have several rows for your table, study the data and write an equation expressing the derivative *in terms of the y-value.* (If necessary, develop more rows of the table.)

2. Repeat the process described in Question 1, but this time use the function whose equation is $y = 10^x$.

3. Pick a third exponential function and go through the process from Question 1 for that function.

HOMEWORK 19

The Forgotten Account

Tyler and his friends (from *Homework 3: Story Sketches*) had \$50 left over after buying uniforms. They decided to put the money into a bank account so the next year's team would have a head start.

Well, next year's team forgot about the account, and so did the following year's team, and pretty soon, the account was completely forgotten.

Suppose the account earned 4.5% interest at the end of each year. (For simplicity, assume that the account was opened on January 1.) The interest is always added to the amount already in the account, so each year's interest is on a larger amount than the year before. (This is called *compound interest*. More specifically, the interest for this account is said to be *compounded annually* because interest is added to the account at the end of each year.)

1. How much money was in the account five years after Tyler and his friends started it?
2. Find a formula that describes the amount of money in the account after t years.
3. Write an equation that could be used to figure out how many years it would take for the account to grow to \$500.
4. Solve your equation from Question 3.
5. Express the solution to the equation from Question 3 as a logarithm.

HOMEWORK 20

How Does It Grow?

To solve the central unit problem, you will need to find a function that grows in the same way that populations grow.

Question 1 of this assignment looks at a simple situation in order to establish some intuitive sense of what to expect from population growth.

1. Suppose a town has a population of 5000 and the population grows by 40 people in a single year. How much growth would you expect for a similar town with a population of 10,000? Explain your reasoning.

Questions 2 through 5 present two specific functions as possible models to describe population as a function of time. You will begin by getting some data about each function and its derivative.

2. First, consider the linear function $f(x) = 3x + 4$.
   a. Choose three different values for x and find $f(x)$ for each of these values.
   b. Find the derivative of the function at each of your three x-values.

Continued on next page

3. Next, consider the quadratic function $g(x) = x^2 - 9$.

   a. Choose three different values for x and find the value of the function for each of these values.

   b. Find the derivative of the function at each of your three x-values.

In Questions 4 and 5, remember that the derivative describes a rate of growth.

4. Now return to the function $f(x) = 3x + 4$ from Question 2.

   a. Describe any relationships you see between the derivative and either x or $f(x)$.

   b. Explain why the relationships you found in Question 4a are or are not appropriate to use in a mathematical representation of population growth. Consider whether those relationships seem to hold for the situation in Question 1.

5. Finally, look at the function $g(x) = x^2 - 9$ from Question 3.

   a. Describe any relationships you see between the derivative and either x or $g(x)$.

   b. Explain why the relationships you found in Question 5a are or are not appropriate to use in a mathematical representation of population growth. Consider whether those relationships seem to hold for the situation in Question 1.

HOMEWORK 21

The Significance of a Sign

When you are graphing a function, it's helpful to know the signs of the coordinates of the points. For instance, knowing the signs of the coordinates tells you what quadrant a point is in, and if a coordinate is 0, you know that the point is on a coordinate axis.

In this assignment, you'll explore similar issues concerning the sign of the derivative of the function.

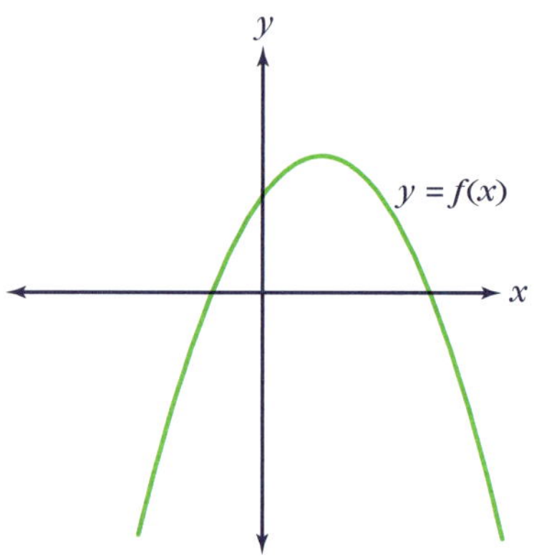

1. Make a copy of the graph at the left. Then identify where on the graph the function's derivative is positive, where the derivative is negative, and where the derivative is 0. (*Hint:* The derivative at a point on the graph can be thought of as the slope of the tangent line at that point.)
2. Sketch the graph of a function for which the derivative is positive for all values of x.
3. Sketch the graph of a function for which there are exactly two points where the derivative is 0.

HOMEWORK 22

The Sound of a Logarithm

Logarithms are a convenient concept for talking about exponents. For instance, the expression $\log_{10} 564$ represents the solution to the equation $10^x = 564$.

Logarithms are also used as the basis for certain units of measurement in science and are particularly useful when we are mainly concerned with changes in order of magnitude.

One such application concerns the measurement of noise levels. The *decibel scale* uses logarithms to describe the intensity of a sound in comparison to a particular reference value. The symbol I_0 is used to represent the intensity of a sound at the threshold of human hearing (that is, the quietest sound that humans can hear).

If I is the intensity of some other sound, the ratio of this intensity to I_0 is called the *relative intensity* of that other sound. We will represent this ratio $\frac{I}{I_0}$ by the letter R. Using this ratio, the *noise level* of that other sound, represented by N, is given by the equation

$$N = 10 \log_{10} R$$

The value of N is given in *decibels,* abbreviated as dB.

For example, suppose a sound has an intensity I that is 100 times the intensity of a sound at the threshold of human hearing. That is, suppose that the relative intensity R is 100. Because $\log_{10} 100 = 2$, this means that the sound has a noise level of $10 \cdot 2$, or 20 decibels.

Continued on next page

A sound at the threshold of human hearing, for which I is equal to I_0, has a noise level of 0 decibels, because $R = 1$, and $10 \cdot \log_{10} 1 = 10 \cdot 0 = 0$.

1. A normal conversation has a relative intensity of approximately 100,000. What is its decibel level?

2. A loud police whistle has a decibel level of approximately 90.

   a. How does its intensity compare to the threshold of human hearing?

   b. How does its intensity compare to ordinary conversation?

3. A sound intensity of approximately 10^{13} times the threshold level will cause a person's ears to hurt. What is the decibel level of such a sound?

4. a. If one sound measures 42 decibels, and the relative intensity of a second sound is three times the relative intensity of the first, what is the decibel level of the second sound?

   b. If one sound measures 60 decibels and another measures 68 decibels, how does the relative intensity of the second sound compare to the relative intensity of the first sound?

CLASSWORK

The Power of Powers

You saw in *Slippery Slopes* that certain functions have the special property that the derivative at each point is proportional to the y-value at that point.

For instance, in the case of the function $f(x) = 2^x$, you found that an equation very much like

$$f'(x) = 0.69 \cdot f(x)$$

appeared to hold true for all values of x. In the case of the function $g(x) = 10^x$, the equation was approximately

$$g'(x) = 2.30 \cdot g(x)$$

The Proportionality Property

When such a relationship between the derivative and the y-value holds true, we say that the function has the "proportionality property." (This is not standard terminology. We are using it here as shorthand for a complex idea.)

Continued on next page

When a function has the proportionality property, the ratio between the derivative and the y-value is called the **proportionality constant.** For instance, the numbers 0.69 and 2.30 are (approximately) the proportionality constants for the functions f and g.

Because the growth of a population is often proportional to the population itself, the proportionality property of functions like f and g makes them excellent candidates for a mathematical model of population growth.

In this assignment, you begin an exploration of which functions have the proportionality property. The exploration continues in *Homework 23: The Power of Powers, Continued.*

The Examples

For each of the two functions shown here, go through these steps.

- Estimate the value of the derivative for at least three different points on the graph.
- Based on your results, state whether you think the function has the proportionality property.
- If you think the function has the proportionality property, give an approximate value for the proportionality constant.

1. $h(x) = 2^x + 3^x$

2. $p(x) = 4.6 \cdot 2^x$

The Power of Powers, Continued

In *The Power of Powers,* you examined whether two particular functions had the proportionality property. This assignment continues the investigation of this property.

1. For each of the three functions shown here, go through these steps.

- Estimate the value of the derivative for at least three different points on the graph.
- Based on your results, state whether you think the function has the proportionality property.
- If you think the function has the proportionality property, give an approximate value for the proportionality constant.

a. $m(x) = 100 + 2^x$

b. $k(x) = 0.83^x$

c. $n(x) = 2^{3x}$

2. Based on your answers in Question 1 and your results from *The Power of Powers,* make some conjectures about the general form of functions that have this special property.

DAYS 24–29

The Best Base

In their work with bases and exponents, Nikki Robinson and Ameilia Mitcalf discuss which might be the "best" base.

You've seen that exponential functions might be an excellent choice for modeling population growth. But what base should you use? 2? 10? Does it matter?

This portion of the unit begins with an activity in which you'll investigate whether bases are interchangeable. But with regard to derivatives, there's one that's clearly the best choice. And you'll see that this special base has a connection with a concept that at first seems completely unrelated.

A Basis for Disguise

You've seen that the derivative of an exponential function is a fixed multiple of the function's *y*-value.

Another useful fact is that exponential expressions can sometimes "disguise" themselves in other number bases. For example, 81^5 can also be written as 3^{20}. Here, a power of 81 is "disguised" as a power of 3. In this activity, you will explore this change-of-base idea.

1. How can you show that 81^5 and 3^{20} are equal without finding the numerical value of either expression? (*Hint:* What special relationship is there between 81 and 3?)

Continued on next page

2. Find a general rule for writing powers of 81 as powers of 3. That is, imagine that you want to put something in the box to fit the equation $81^x = 3^{\square}$. Explain how the number that goes in the box depends on x. (*Suggestion:* Think about how much base 3 cake would have the same effect on Alice as x ounces of base 81 cake.)

3. Reverse the roles of the two bases, 81 and 3, and find a general rule for writing powers of 3 as powers of 81. (The Alice metaphor may help here, too.)

4. Questions 1 through 3 may seem like special cases, because 81 is a whole-number power of 3.

   a. Suppose the two bases are 7 and 5. Can you find a general rule for writing 7^x as a power of 5? (*Hint:* Can you write 7 itself as a power of 5? Does the Alice metaphor help?)

   b. Examine whether there is always a general rule for writing b^x as a power of a, no matter what numbers are used for a and b. (Of course, the rule itself would have to depend on a and b.)

   Give the rule in the cases for which such a rule exists, and describe the values for a or b for which the rule would not exist.

HOMEWORK 24

Blue Book

Car dealers sometimes use the rule of thumb that a car loses about 30% of its value each year. Use this rule to answer Questions 1 and 2.

1. Suppose you bought a new car in December 1990 for $15,000.

a. According to the rule of thumb, what would the car be worth at each of these times?

i. In December 1991

ii. In December 1995

iii. In December 2000

b. Develop a general formula for the value of the car *t* years after purchase.

2. Clarabell has thought about how cars lose value. She noticed that a $20,000 car will lose about $6,000 of its value the first year, while a $10,000 car will lose about $3,000 of its value the first year. She figures that the more expensive car loses more value each year, so eventually it will be worth less than the cheaper car.

How long do you think it will take until this happens?

HOMEWORK 25

California and Exponents

In 1850, the population of California was about 92,600, and in 1860, it was about 380,000. In *Homework 8: California, Here I Come!,* you found a linear function that went through the two points (1850, 92,600) and (1860, 380,000).

Your task in this assignment is to find an exponential function for the same population data and then examine its accuracy for making predictions. To make the arithmetic much simpler, treat the year 1850 as $x = 0$ and the year 1860 as $x = 10$.

1. Find two numbers a and b so that the exponential function $y = a \cdot b^x$ goes through the two points (0, 92,600) and (10, 380,000). (*Hint:* Substitute the coordinates of the first point for x and y to get an equation that involves only a. Solve that equation and then use the second point and the value you found for a to find b.)

2. Use your answer to Question 1 to determine what the population of California would have been in 1990 if population growth had continued at the same relative growth rate as during the Gold Rush period. (*Hint:* If the value 0 for x represents 1850 and the value 10 for x represents 1860, what value of x represents 1990?)

3. Based on this exponential model, do you think population growth in California has slowed down or speeded up since the Gold Rush period?

Find That Base!

You know that the derivative of the exponential function $y = b^x$ is a particular constant times its y-value. As long as the base b is fixed, this proportionality constant is the same at every point on the graph, but the proportionality constant does depend on the value used for b.

Scientists prefer to pick a standard base so it will be easier to compare one function to another. Because any exponential function can be expressed in any base (as long as you stick to positive bases other than 1), they are free to pick any number for this standard base.

Because scientific work often involves derivatives, scientists have chosen for this standard base the number that makes the proportionality constant equal to 1.

Your job is to estimate this special base. In other words, estimate the value of b for which, at every point on the graph of the function $f(x) = b^x$, the derivative is equal to b^x.

HOMEWORK 26

Double Trouble

In inflation situations like that of *How Much for Broken Eggs?!!?*, the length of time it takes for the price to double is called the *doubling time*.

Of course, the doubling time depends on the rate of inflation. Your main task in this assignment is to investigate the relationship between these two values.

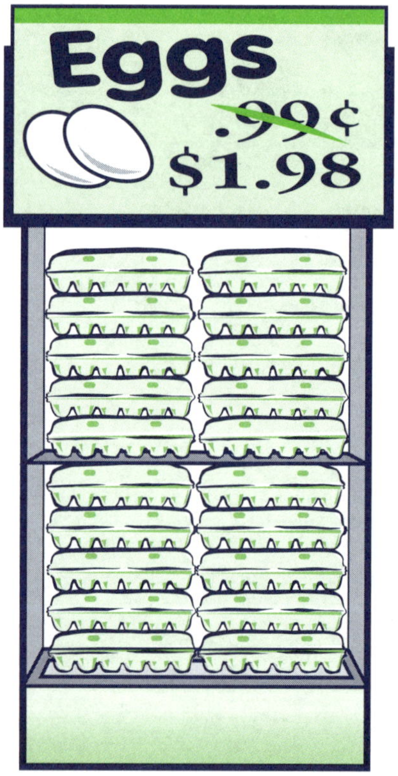

1. Consider the case of an inflation rate of 5% per year, and find its doubling time. For simplicity, you can start with a price of \$1 and see how long it takes until the price gets to \$2. Give your answer to the nearest hundredth of a year.

2. Choose a different inflation rate, and find the doubling time.

3. Continue with other inflation rates, and put your results in an In-Out table in which the *In* is the inflation rate and the *Out* is the doubling time.

4. a. Look for a pattern or rule that describes your In-Out table.

   b. Try to explain why your pattern works.

5. a. For each inflation rate in your table, also compute the *quadrupling time*—that is, the length of time it takes for the price to be multiplied by a factor of 4. (If you started with \$1, this would be the time required to reach \$4.)

   b. How does the quadrupling time compare with the doubling time? Explain your answer.

The Generous Banker

"Double Your Money in 20 Years!" read the bank's advertisement. Adam thought that sounded like a pretty good deal, but he also thought he might be able to talk his way into something even better. So he went in to speak to the banker.

"Doubling your money is like increasing it by 100%, I think," he said with a practiced uncertainty. "But what if I need my money before 20 years? Can I get a proportional part of the interest each year, just in case?"

The banker hesitated, never having studied much mathematics. "Well, that seems fair, Mr. Smith," she finally replied. "How much should you get each year?"

Adam feigned confusion and took out his calculator. "100% for 20 years . . . let's see . . . , 100 divided by 20 . . . , I guess that's 5% each year. Can you increase my account by 5% each year instead?"

The banker agreed, and so Adam deposited $1,000 in an account and left it there for the full 20 years.

1. How much was in the account 20 years later?
2. How much would have been in the account if Adam had talked the bank into giving him a proportional amount of interest, compounded every six months instead of once a year?
3. What if Adam got a proportional amount of interest, compounded every three months?

HOMEWORK 27

Comparing Derivatives

Part I: Shared Points

Here are the equations for three functions whose graphs all pass through (0, 0). The three graphs also all pass through (1, 1).

- $f(x) = x$
- $g(x) = x^2$
- $h(x) = x^3$

Do you think all three functions will have the same derivative at (0, 0)? What about at (1, 1)? The following questions will help you decide.

1. Graph all three functions on the same set of axes for x-values from -1 to 2. Be sure to plot enough specific points for each function (including noninteger values of x) to get accurate graphs, and pay particular care in plotting x-values between 0 and 1.

2. *Based only on your graphs,* answer each of these questions and explain your reasoning.

   a. Which of the three functions has the largest derivative at the point (0, 0)?

   b. Which of the three functions has the largest derivative at the point (1, 1)?

3. Actually find the derivative of each function at each of the common points, and compare your results with your answers to Question 2.

Continued on next page

Part II: Derivative Sketch

Suppose this is the graph of the function defined by the equation $y = f(x)$.

4. Make a copy of the graph, and show where on the graph the derivative of this function is positive, where the derivative is negative, and where the derivative is 0. Make your answers as complete as possible.

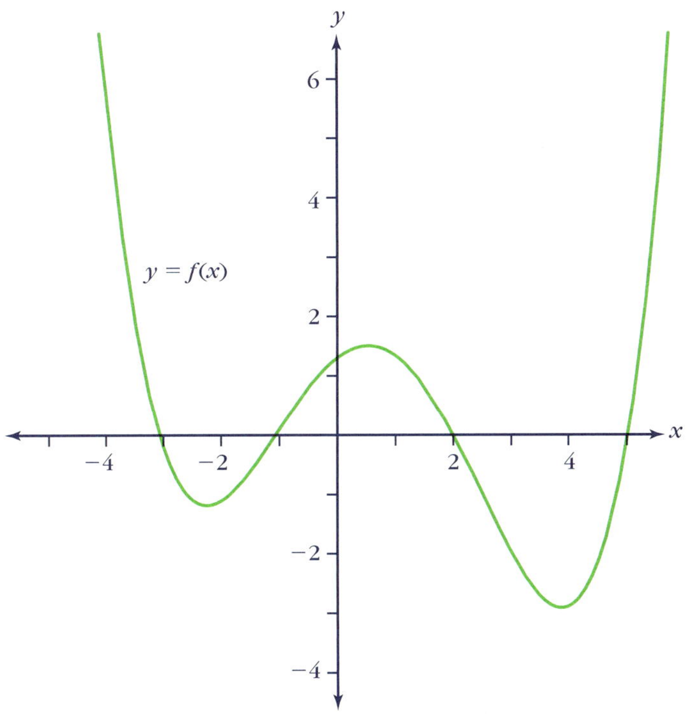

5. *Sketch* the graph of f', which is the derivative function of f. Use the same scale for the x-axis on your graph of the derivative function as you use for the graph of f itself.

The Limit of Their Generosity

As you saw in *The Generous Banker,* Adam deposited $1,000 in a bank. The bank had advertised that it would double his money in 20 years. But Adam persuaded the banker to increase it instead by 5% for each of the 20 years instead of increasing it by 100% all at once at the end of the 20-year term.

For a while, Adam thought he was going to get really rich due to the banker's generosity. He persuaded her to give him an appropriate fraction of the interest every six months and then persuaded her to calculate the interest every month.

Once he had gotten this far, Adam was ready to spring his ultimate request. He asked the banker to give him proportional interest every day and add it to his account.

1. Without doing any calculations, about how much do you think Adam would end up with?
2. Using your calculator, compute Adam's wealth after 20 years. (Figure 365.25 days per year.)
3. What would Adam end up with if the bank gave him proportional interest *every hour*?

California Population with e's

In *Homework 25: California and Exponents,* you used a function of the form $y = a \cdot b^x$ to get a model of the population growth of California between 1850 and 1860.

Specifically, using $x = 0$ to represent 1850 and $x = 10$ to represent 1860, you found the values of a and b so that the function $y = a \cdot b^x$ went through the points (0, 92,600) and (10, 380,000).

Your task in this assignment is similar, but now you should use e as the base of the exponential function.

1. Find values for k and c so that the exponential function $y = k \cdot e^{cx}$ goes through those two points. (*Hint:* As in the previous assignment, use the first point to get an equation involving only one of the two unknown values.)

2. Express the value you get for c in terms of a natural logarithm.

3. How are the coefficients k and c in this assignment related to the coefficients a and b in the previous assignment? (If necessary, redo Question 1 of *Homework 25: California and Exponents* to find a and b.)

DAYS 30–32

Back to the Data

Which type of function is Joe Stebbins using to predict how long before the population will read $1.6 \cdot 10^{15}$?

You're now ready to reexamine the population data that you saw at the beginning of this unit. As a first step, you'll explore how to "tweak" a function—that is, how to make little changes in the function so that its graph is closer to what you want.

You'll then conclude the unit by looking for a function that comes close to the initial data and using that function to predict when we will be squashed up against one another.

Tweaking the Function

Suppose that you have some data points and need to fit them with a function. Suppose also that you know more or less what type of function should fit the data and that your first guess comes pretty close to fitting, but not as close as you would like.

You then need to change the function a bit—to adjust it somehow to make it fit the data somewhat better.

How do you do this? This activity should help you answer that question.

Begin with the function $y = e^x$. Change it in different ways and watch how the graph is affected. For instance, you might try multiplying e^x by various coefficients, or you might adjust the exponent in some way.

Your task is to explore how the graph is changed. Here are some things to watch for.

- What makes the graph "more curvy" or "less curvy"?
- What changes the horizontal position or the vertical position of the graph?

Begin with these questions, but do not limit yourself to them. Find out whatever you can and keep track of what you learn.

HOMEWORK 30

Beginning Portfolios—Part I

Think about the examples of growth and change that you have studied in this unit, focusing on two particular kinds of functions.

- Linear functions, which have the form $y = a + bx$
- Exponential functions, which have the form $y = k \cdot e^{cx}$

Compare the two types of functions, addressing these issues.

- How does each function represent rates of growth or change?
- How does each function represent starting points or initial values?
- What kinds of situations is each function appropriate for describing or modeling? (Be more specific than simply saying "linear growth" or "exponential growth.")

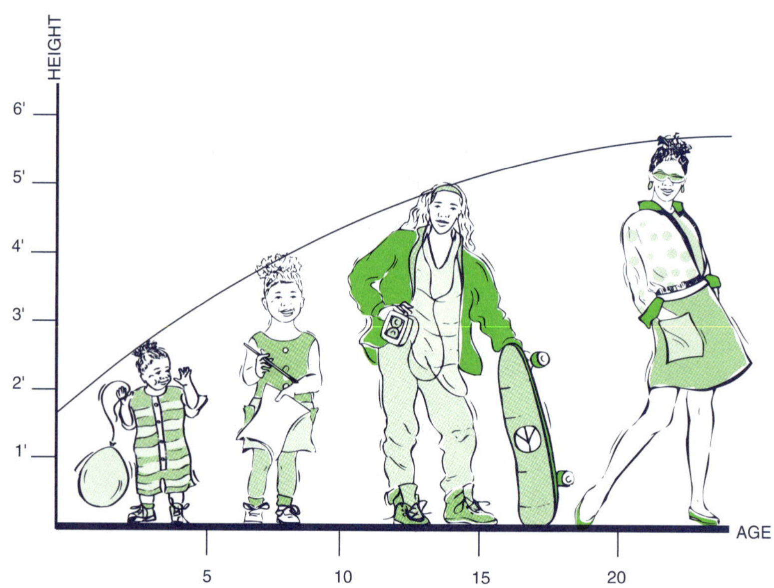

Return to *Small World, Isn't It?*

Here are the numbers you saw at the beginning of the unit, which show the changes in world population over the past several centuries.

Year	Estimated Population
1650	470,000,000
1750	694,000,000
1850	1,091,000,000
1900	1,570,000,000
1950	2,510,000,000
1960	3,030,000,000
1970	3,680,000,000
1980	4,480,000,000
1985	4,870,000,000
1990	5,290,000,000
1995	5,730,000,000

As initially stated, your task was to determine, based on this data, how long it would take until people were "squashed up against one another." In this unit, that phrase has been interpreted to mean that each person has exactly 1 square foot to call her or his own. Based on estimates of the earth's surface area, this means that the population would have to reach approximately $1.6 \cdot 10^{15}$ people.

Continued on next page

Here are your tasks in this final activity.

1. Plot the population data.

2. a. Find a function that approximates your data. At least at first, you should look for a function of the form $y = k \cdot e^{cx}$. This first approximation need not be very accurate.

   b. Assuming that the population grew according to your function from step a, determine when each person would have only 1 square foot to call his or her own. That is, based on the function from step a, how long would it take for the population to reach $1.6 \cdot 10^{15}$ people?

3. Repeat steps a and b from Question 2 as often as you think is useful, looking for a better approximation. Use ideas from the activity *Tweaking the Function* to try to make whatever adjustments you think are needed.

   State your best approximating function and the estimate it gives for how long it would take for the population to reach $1.6 \cdot 10^{15}$ people.

4. Discuss whether you think the population data in the table is really exponential and whether you think it will be exponential in the future. Give reasons for your conclusions.

HOMEWORK 31

Beginning Portfolios—Part II

1. Much of this unit focused on exponential functions, which leads naturally into working with compound interest.

   Pick an activity from the unit that helped develop your understanding of compound interest, and describe what you learned from the assignment.

2. The beginning of this unit concentrated on linear functions. Summarize what you have learned about finding equations of straight lines, and select one or two activities that were particularly helpful to you in this area.

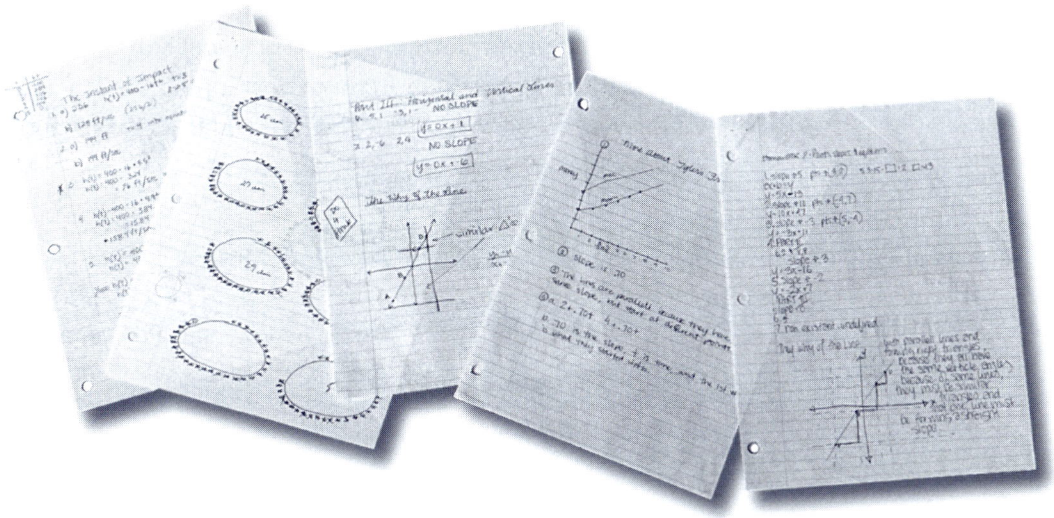

HOMEWORK 32

Small World, Isn't It? Portfolio

Now that *Small World, Isn't It?* is completed, it is time to put together your portfolio for the unit. Compiling this portfolio has three parts.

- Writing a cover letter in which you summarize the unit
- Choosing papers to include from your work in this unit
- Discussing your personal growth during the unit

Cover Letter for *Small World, Isn't It?*

Look back over *Small World, Isn't It?* and describe the central problem of the unit and the main mathematical ideas. This description should give an overview of how the key ideas were developed and how they were used to solve the central problem.

In compiling your portfolio, you will select some activities that you think were important in developing the key ideas of this unit. Your cover letter should include an explanation of why you selected the particular items.

Selecting Papers from *Small World, Isn't It?*

Your portfolio for *Small World, Isn't It?* should contain these items.

- *Homework 16: What's It All About?*
- *Homework 30: Beginning Portfolios—Part I*
- *Homework 31: Beginning Portfolios—Part II*

  Include the activities from the unit that you selected as part of this assignment.

Continued on next page

- Other key activities

  Include two or three other activities that you think were important in developing the key ideas of this unit.

- A Problem of the Week

  Select one of the POWs you completed during this unit (*The More the Merrier?*, *Planning the Platforms*, or *Around King Arthur's Table*).

Personal Growth

Small World, Isn't It? focused on various kinds of growth. For this part of your portfolio, write about your own growth during the unit. You might want to specifically address this issue.

> *What changes or growth have you noticed during Year 3 in your ability to work well in groups?*

You should include here any other thoughts about your experiences with this unit that you want to share with a reader of your portfolio.

Appendix

Supplemental Problems

The supplemental problems for *Small World, Isn't It?* continue the unit's areas of emphasis—rates of change and linear and exponential functions—although other topics appear as well. Here are some examples.

- *Solving for Slope* and *The Slope's the Thing* give additional perspectives on how to find the slope for a linear equation.
- *Deriving Derivatives* looks at developing general formulas for derivatives.
- *Dr. Doubleday's Base* and *Investigating Constants* examine the role of the parameters in two different forms of the general exponential function.
- *Summing the Sequences—Parts I and II* are follow-ups to *POW 9: Planning the Platforms.*

SUPPLEMENTAL PROBLEM

Solving for Slope

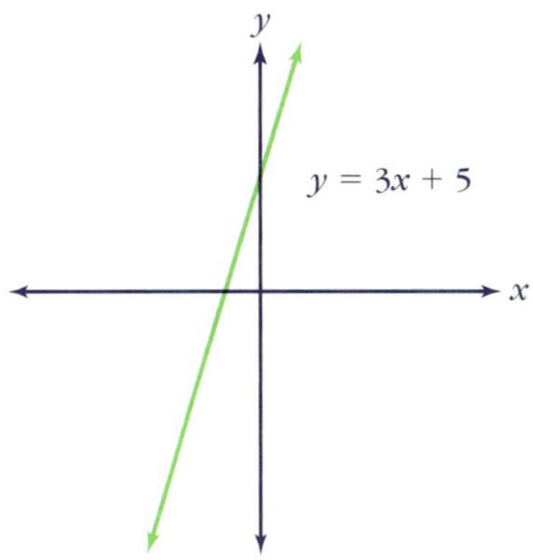

You've seen that in an equation like $y = 3x + 5$, the coefficient of x (in this case, 3) is the slope of the graph.

In this activity, you will examine linear equations that do not have this form.

1. Begin by examining these specific examples. In each case, solve the equation for y in terms of x to determine the slope.

   a. $3y - 2x = 12$

   b. $x + 4y = -8$

   c. $5x - 2y = 10$

   d. $-6y + 7x = -2$

2. Now consider the general linear equation in the standard form

$$Ax + By = C$$

   a. Develop an expression for the slope in terms of A, B, and C.

   b. Consider each of these questions. Based on examples, describe how to answer them simply by looking at A, B, and C.

      - Is the line rising or falling as it goes to the right?
      - For lines rising to the right, is the line steeper than $y = x$?
      - For lines falling to the right, is the line steeper than $y = -x$?

      Explain not only how your method works, but also why it works.

SUPPLEMENTAL PROBLEM

Slope and Slant

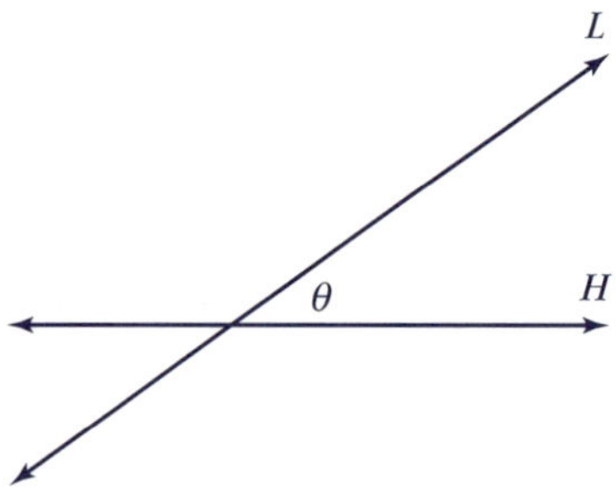

One way to describe the "steepness" or "slant" of a line is to measure the angle formed between that line and a horizontal line. This angle is called the **angle of inclination** of the line.

For instance, the diagram here shows a line L and a horizontal line H. The angle labeled θ (the Greek letter "theta") is the angle of inclination of line L.

You've been working with lines that are the graphs of linear equations. In the context of the coordinate system, the angle of inclination of a line is the angle formed between that line and the positive direction of the x-axis. You might expect to find some relationship between the slope of such a line and its angle of inclination.

Finding such a relationship is complicated by the fact that the steepness of the graph is affected by the scales on the axes. In this activity, you will eliminate that complication by assuming that the vertical and horizontal axes have the same scale.

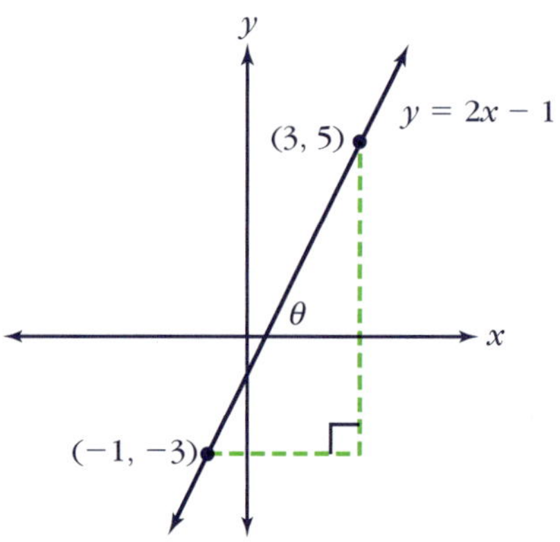

1. This diagram shows the graph of the equation $y = 2x - 1$ and two points, $(-1, -3)$ and $(3, 5)$, that fit this equation. The diagram also shows an angle of inclination θ for the line and a right triangle with $(-1, -3)$ and $(3, 5)$ as two of its vertices.

   a. Find the slope of the line $y = 2x - 1$.

   b. Use trigonometry and the right triangle to find a relationship between the slope of this line and its angle of inclination.

Continued on next page

c. Use your work from part b to find the measure of angle θ.

d. Carefully draw the line through $(-1, -3)$ and $(3, 5)$ on graph paper, using the same scale for the vertical and horizontal axes. Then measure the angle of inclination and check if it matches your result from part c.

2. Use the ideas from Question 1 to answer these questions (again making the assumption that the vertical and horizontal axes have the same scale).

   a. What is the angle of inclination for a line whose slope is 1?

   b. What is the slope of a line whose angle of inclination is 30 degrees?

3. State a general principle, using trigonometry, relating the slope of the line to its angle of inclination.

   Note: So far, we have defined the trigonometric functions only for acute angles in right triangles. Therefore, in Question 3, assume that the line has an angle of inclination between 0 and 90 degrees. (You will see how to define the trigonometric functions for arbitrary angles in the Year 4 unit *High Dive.*)

SUPPLEMENTAL
PROBLEM

Predicting Parallels

When a linear equation is written to express y in terms of x, the coefficient of x is equal to the slope of the graph. For example, the graph of $y = 3x + 5$ has slope 3. This makes it easy to recognize when two equations have graphs that are parallel, because they have the same coefficient for x.

According to this principle, the graphs of the equations $y = 3x + 5$ and $y = 3x + 9$ should be parallel. That is, they should not have any points in common. The first task in this activity is to prove this fact.

1. Show that the equations $y = 3x + 5$ and $y = 3x + 9$ cannot have any common solutions. That is, show that there is no point that is on the graphs of both equations.

But what about lines that are not in "$y =$" form? Is there a simple way to recognize such lines in the standard form $Ax + By = C$ are parallel? Begin by answering Questions 2 and 3, and then try to generalize your results.

2. a. Draw the graph of the equation $4x - 3y = 24$.
   b. Choose a point *not* on the graph, and draw a line through that point parallel to your graph from part a.
   c. Find the equation of the line you drew in part b.

3. Graph the equation $8x - 6y = 24$, and compare it with the graphs from Question 2.

The Generalization

4. Based on Questions 2 and 3, and from other equations you might examine, what general principles can you state for determining at a glance whether two linear equations in standard form have parallel graphs?

SUPPLEMENTAL PROBLEM

The Slope's the Thing

You know that two points determine a straight line, so knowing the coordinates of two points on a line is enough information to get an equation for that line. This activity looks at a systematic way to get the equation from that information, based on slope.

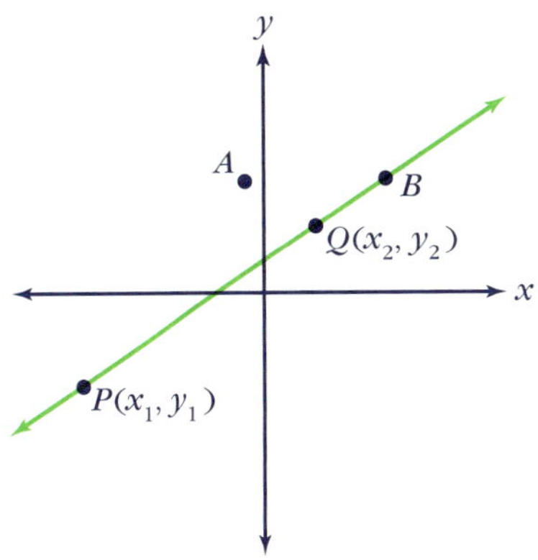

Let's call the two given points P and Q. The method is based on these two statements.

- If a point is *not* on that line (for instance, point A in the diagram), then the slope of the line connecting that point to P will be *different* from the slope of the line through P and Q.
- If a point *is* on that line (for instance, point B), then the slope of the line connecting that point to P will be *the same* as the slope of the line through P and Q.

1. Suppose that P has coordinates (x_1, y_1), that Q has coordinates (x_2, y_2), and that R is another point in the plane, with coordinates (x, y).

  a. Write an expression for the slope of the line through P and Q.

  b. Write an expression for the slope of the line through P and R.

  c. Combine parts a and b to get an equation stating that these two slopes are equal.

2. Apply this method to find the equation of the line through $(1, 5)$ and $(-2, -1)$.

3. Use similar triangles to prove the two statements on which this method is based.

SUPPLEMENTAL
PROBLEM

Speedy's Speed by Algebra

In *Photo Finish,* you met your old friend Speedy the track star, who runs the last 400 meters of a 1600-meter relay race.

Based on a film of one of her races, an analyst came up with the function $m(t) = 0.1t^2 + 3t$ to describe how far Speedy had run after t seconds of her segment of the race. The equation $m(50) = 400$ shows that Speedy crossed the finish line after exactly 50 seconds.

In *Photo Finish,* you found Speedy's average speed for small time intervals near the end of the race and used those average speeds to get an estimate for her speed at the instant when she crossed the finish line. You may wonder how accurate or reliable that estimate is. That is, you may be wondering this:

> *How fast was Speedy really going at the instant when she crossed the finish line?*

This assignment describes an algebraic approach to answering this question.

1. First, use steps a through c to write out in detail the computation involved in finding Speedy's average speed for the last tenth of a second at the end of the race.
    a. Find out how far Speedy had run after 49.9 seconds. That is, compute $m(49.9)$.
    b. Find out how far Speedy ran during the last tenth of a second. That is, find the difference between $m(49.9)$ and 400.

Continued on next page

c. Find Speedy's average speed during the last tenth of a second of the race. That is, divide the distance you found in Question 1b by the length of the time interval, which is 0.1 seconds.

2. Now use the same process as in Question 1, but substitute the last h seconds for the last tenth of a second. That is, instead of working with the time interval from $t = 49.9$ to $t = 50$, as in Question 1, work with the time from $t = 50 - h$ to $t = 50$. Steps 2a through 2c will guide you through this process.

   a. Find out how far Speedy had run after $50 - h$ seconds. That is, find $m(50 - h)$. (Your answer should be an expression in terms of h.)

   b. Find out how far Speedy ran during the last h seconds. That is, write an expression (in terms of h) for the difference between $m(50 - h)$ and 400.

   c. Divide the distance you found in Question 2b by the length of the time interval, which is h seconds.

   d. Confirm your expression in Question 2c by substituting 0.1 for h. Does that give the same answer you got in Question 1c?

3. Your result for Question 2c should be an algebraic expression in terms of h. This expression gives Speedy's average speed during the last h seconds of the race.

   a. Simplify this expression as much as you can.

   b. What happens to your expression in Question 3a as h gets smaller and smaller?

   c. What does your result in Question 3b mean in terms of Speedy's instantaneous speed?

4. Does your work in Questions 1 through 3 give you further confidence in your results from *Photo Finish*? Explain your response.

SUPPLEMENTAL
PROBLEM

Potential Disaster

Construction workers have partially ruptured a natural gas pipeline in a research building. The pipeline's protective inner membrane has been pushed out through the opening, forming a sphere-shaped balloon on the outside of the pipeline.

Emergency crews are confident that they can patch the hole in the pipe but see no way to push the balloon back into the pipe. They decide that they must cut the balloon off from the pipe and then quickly patch the hole.

An electrical wire nearby poses a serious danger, because if the wire and balloon come in contact with each other, the gas in the balloon will be ignited and the balloon will explode, causing massive damage. Gas is flowing into the balloon so that its volume is growing at a constant rate.

The balloon was discovered 30 minutes after the rupture. In those 30 minutes, the balloon had grown to a diameter of 1 foot. The balloon will come in contact with the wire if its diameter reaches 2 feet.

It is now 20 minutes since the balloon was discovered, and the building's manager is in a panic. The manager thinks that the balloon will explode in only 10 more minutes. But the engineer in charge of the emergency crew seems rather calm.

1. a. Why do you suppose the manager thinks they have only 10 minutes left before the balloon explodes?
   b. What is wrong with the manager's reasoning?
2. How much time do they really have? (The volume of a sphere of diameter d is given by the expression $\frac{1}{6}\pi d^3$.)

Proving the Tangent

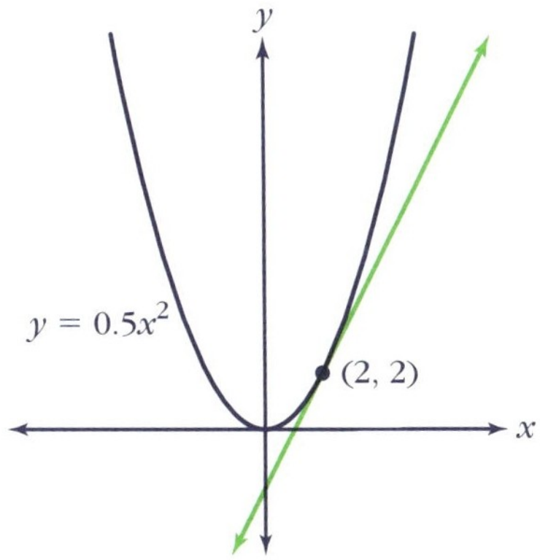

In *Homework 15: On a Tangent,* you examined the graph of the function $f(x) = 0.5x^2$ near the point (2, 2) on the graph. You were asked to use a series of secant lines through (2, 2) to estimate the slope of the tangent line at that point, which seemed to be equal to 2.

1. Use this estimate of the derivative to write the equation of the tangent line. That is, write the equation of the line whose slope is 2 and that goes through the point (2, 2).
2. Prove that your estimate of the derivative is correct by showing that (2, 2) is the only point where the line you found in Question 1 meets the graph of the function $f(x) = 0.5x^2$. (*Hint:* Show that your equation from Question 1 and the equation $y = 0.5x^2$ have only one solution in common.)

SUPPLEMENTAL
PROBLEM

Summing the Sequences—Part I

In *POW 9: Planning the Platform,* Camilla needed to find the sum of a sequence of numbers in which each term differed from the previous term by the same amount.

For instance, suppose that the first platform had been 28 inches tall, that the difference in height between adjacent platforms had been 8 inches, and that there had been 5 platforms altogether. Then she would have needed to find the sum

$$28 + 36 + 44 + 52 + 60$$

A sequence such as 28, 36, 44, and so on, in which the difference between terms is constant, is called an **arithmetic sequence.** (In this context, the word "arithmetic" is an adjective and is pronounced "a-rith-*met*-ic," with the emphasis on the third syllable.)

The first number in an arithmetic sequence is called the **initial term,** and the amount added to get each successive term is called the **difference.** For instance, in our example, the initial term is 28 and the difference is 8.

1. a. Find an expression in terms of n for the nth term of the arithmetic sequence whose initial term is 28 and whose difference is 8.

   b. Find an expression in terms of n for the sum of the first n terms of this sequence.

Continued on next page

If we use a to represent the initial term and d to represent the difference, then the general arithmetic sequence has the terms a, $a + d$, $a + 2d$, $a + 3d$, and so on.

2. Find an expression in terms of a, d, and n for the nth term of the general arithmetic sequence.

3. Find an expression in terms of a, d, and n for the sum of the first n terms of the general arithmetic sequence. Give your answer as an algebraic expression in closed form—that is, without the use of summation or ellipsis (. . .) notation.

4. Apply your expression from Question 3 to find the sum of the first 50 terms of the arithmetic sequence 15, 21, 27, 33, and so on.

SUPPLEMENTAL
PROBLEM

Summing the Sequences—Part II

In *Summing the Sequences—Part I,* you examined arithmetic sequences, which have a constant difference between terms.

If instead we require a constant *ratio* between terms, we get another important category of sequences, the **geometric sequence.** For instance, the sequence 8, 24, 72, 216, and so on, in which each term is three times the previous term, is an example of a geometric sequence. Here, 8 is the initial term and 3 is the ratio.

1. a. Find an expression in terms of n for the nth term of the geometric sequence whose initial term is 8 and whose ratio is 3.

   b. Find an expression in terms of n for the sum of the first n terms of this sequence.

   Hint: Compare the sum of the first n terms of this sequence with the sum of the first n terms of the sequence 24, 72, 216, and so on. The second sum is three times the first (because each term is three times as big). What do you get if you subtract the first sum from the second? (Think about terms that cancel out. You may want to look at specific values of n.)

If we use a to represent the initial term and r to represent the ratio, then the general geometric sequence has the terms a, ar, ar^2, ar^3, and so on.

Continued on next page

2. Find an expression in terms of a, r, and n for the nth term of the general geometric sequence.

3. Find an expression in terms of a, r, and n for the sum of the first n terms of the general geometric sequence. Give your answer as an algebraic expression in closed form—that is, without the use of summation or ellipsis (. . .) notation. (Try to generalize the hint from Question 1b.)

4. Apply your expression from Question 3 to find the sum of the first 20 terms of the sequence 3, 6, 12, 24, and so on.

5. Consider the case of the geometric sequence with $a = 1$ and $r = \frac{1}{2}$.

   a. Write an expression (in terms of n) for the sum of the first n terms of this sequence. (Use your work from Question 3.)

   b. Find the sum of the first ten terms of this sequence by actually adding the terms, and use it to verify your answer from Question 5a.

   c. What happens to your expression from Question 5a as n gets bigger? Does that fit your intuitive idea? Explain.

6. (Extra challenge) How can you generalize Question 5 to the case in which the ratio r is any number between 0 and 1?

SUPPLEMENTAL
PROBLEM

A Little Shakes a Lot

One important use of logarithms (at least in some places) is the *Richter scale,* which is a numerical way of describing the magnitude of an earthquake.

Somewhat simplified, the formula used to get the Richter scale number of an earthquake is

$$R = \log_{10} a$$

where R is the Richter scale number and a is the amplitude or amount of the ground motion, as measured on a seismograph.

Continued on next page

In order to give intuitive meaning to Richter scale numbers, you need to have some points of reference. For example, an earthquake that measures 4.0 on the Richter scale is barely perceptible outside its immediate center. In contrast, the great San Francisco earthquake of 1906 measured 8.3 on the Richter scale.

1. These Richter scale numbers make it sound as if the 1906 earthquake was only about twice as big as an earthquake that can hardly be felt. But, in fact, that isn't the case at all. Show this by answering these questions.
   a. How many times as much ground motion does an 8.3 quake have compared to a 4.0 quake?
   b. What Richter scale measurement would represent a quake that has twice the ground motion of one that measures 4.0 on the Richter scale?
   c. What Richter scale measurement would represent a quake that has half the ground motion of one that measures 8.3 on the Richter scale?
2. The 1989 Loma Prieta earthquake in California was measured at about 7.1 on the Richter scale. In numerical terms, how did the amount of its ground motion compare to that of the 1906 earthquake?

SUPPLEMENTAL
PROBLEM

Looking at Logarithms

Logarithms are defined in terms of exponential equations. For instance, $\log_b a$ is defined as the value of x that fits the equation $b^x = a$.

Because of this relationship between logarithms and exponents, there is a principle about logarithms corresponding to every principle about exponents. Your task in this activity is to develop these principles for logarithms. Throughout the activity, b represents a positive number other than 1.

1. The additive law of exponents states

$$b^x \cdot b^y = b^{x+y}$$

Your first task is to find a corresponding principle for logarithms. Start with the specific questions in Questions 1a and 1b.

a. What is $\log_2 2^5$? What is $\log_2 2^9$? What is $\log_2 (2^5 \cdot 2^9)$?

b. Find approximate values for $\log_3 8$, $\log_3 7$, and $\log_3 56$. What relationship do you see among these three logarithms?

c. Generalize your results from Questions 1a and 1b to get a formula for $\log_b (rs)$ in terms of $\log_b r$ and $\log_b s$.

d. Use the additive law of exponents to prove your general result. (*Hint:* Think of r as b^x and s as b^y.)

Continued on next page

2. Another principle for exponents states

$$(b^x)^n = b^{xn}$$

Again, your task is to find a corresponding principle for logarithms by first looking at specific questions.

a. How does $\log_5 (5^7)^6$ compare to $\log_5 5^7$?

b. Find approximate values for $\log_3 17$ and $\log_3 17^4$. What relationship do you see between these two logarithms?

c. Generalize your results from Questions 2a and 2b to get a formula for $\log_b (r^n)$ in terms of $\log_b r$ and n.

d. Use the principle $(b^x)^n = b^{xn}$ to prove your general result. (*Hint:* Again, think of r as b^x.)

3. The range, or set of possible outputs, for the function $y = b^x$ consists of all positive numbers. That is, as x varies over all possible values, y can be any positive number. What does this say about the domain, or set of possible inputs, of the function defined by the equation $f(u) = \log_b u$? That is, what numbers can be used for u in this function?

4. a. What is the domain for the function $y = b^x$?

b. What does your answer to Question 4a say about the range of the function defined by the equation $g(v) = \log_b v$?

SUPPLEMENTAL
PROBLEM

Finding a Function

In *Homework 21: The Significance of a Sign,* you were asked to sketch the graph of one function for which the derivative is positive for all values of x and another function for which the derivative is zero at exactly two points. This activity continues that theme.

1. Sketch the graph of a function f that fits both of these conditions.
   - $f'(x)$ is positive for all values of x between -2 and 3
   - $f'(x)$ is negative if $x < -2$ or $x > 3$
2. Sketch the graph of a function $g(x)$ that fits all of these conditions.
   - $g'(x)$ is zero for exactly three values of x
   - $g(x)$ is positive between $x = -1$ and $x = 4$
   - $g(x)$ is negative if $x > 4$ or $x < -1$

SUPPLEMENTAL PROBLEM

Deriving Derivatives

By looking at the derivative of a function at many of its points, you can sometimes find a rule for its derivative at any point.

In this activity, you will be looking for a rule that allows you to find quickly the derivative of a function of the form $y = ax^n$. (The functions defined by the equations $y = 2x^2$, $y = x^3$, and $y = -3x^6$ are all examples of this type of function.)

Suggestion: Start with a specific function of this form, make a table of values for its derivative, and find a rule for this table. You might then try other functions that use the same exponent and look for a generalization for functions with that exponent. (*Reminder:* This exponent is called the *degree* of the function.)

Once you have a rule for functions of a particular degree, try a different degree and then try to generalize.

SUPPLEMENTAL
PROBLEM

The Reality of Compounding

You saw in *The Generous Banker* that getting 5% annual interest, compounded each year for 20 years, is not the same as doubling your money in 20 years.

1. If interest is compounded annually, what rate of interest should the bank give each year in order for money to double in 20 years?

2. Suppose that the bank compounds interest quarterly. What quarterly rate will double the value of an account in 20 years?

SUPPLEMENTAL PROBLEM

Transcendental Numbers

In *Orchard Hideout,* you worked with the number π, which can be defined as the ratio between the circumference and diameter of a circle. In this unit, you were introduced to the number e. Both π and e are examples of **transcendental numbers.**

What is a transcendental number? Your task in this assignment is to find out about this category of numbers. You may want to start your investigation at the library. Then write a report explaining what you learned about transcendental numbers and their history.

SUPPLEMENTAL PROBLEM

Dr. Doubleday's Base

You have seen that for any positive number b, the derivative of the exponential function $y = b^x$ is a proportionality constant times its y-value.

For example, you found that if the base b is equal to 2, then the proportionality constant is approximately 0.69. In other words, for the function $y = 2^x$, the derivative at a point $(x, 2^x)$ on the graph is approximately $0.69 \cdot 2^x$.

You have also seen that there is a special base, called e, for which this proportionality constant is the number 1. In other words, for the function $y = e^x$, the derivative at any point on the graph is equal to the y-value at that point.

While it is convenient for scientists to use e as the base, Dr. Doubleday would like a base for which the proportionality constant is 2. (We don't know why.) In other words, the doctor would like to find an exponential function whose derivative at every point of its graph is equal to twice its y-value at that point.

1. Determine what base Dr. Doubleday should use. That is, find a number b such that for every point (x, b^x) on the graph of the function $f(x) = b^x$, the derivative is equal to $2 \cdot b^x$. (You will only be able to estimate b. Try to get its value to the nearest hundredth.)
2. Look for a relationship between your answer to Question 1 and the number e.

SUPPLEMENTAL PROBLEM

Investigating Constants

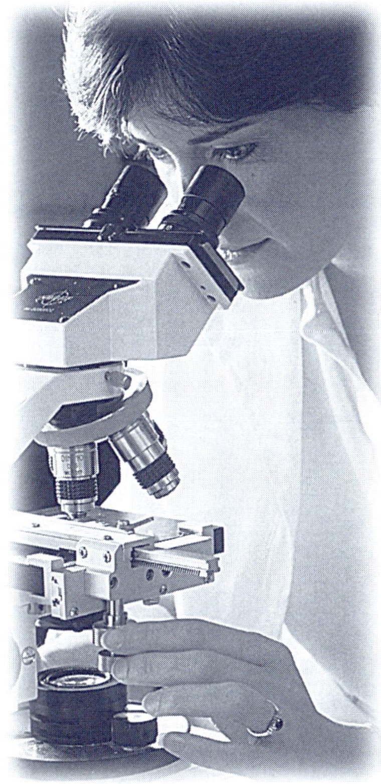

1. You've seen that the general exponential function can be written in the form $y = k \cdot e^{cx}$, where k and c can be any two nonzero numbers. (What happens if k or c is zero?)

   You also know that any such function has the proportionality property. That is, the derivative at any point on the graph is proportional to the y-value at that point.

   Investigate how the value of the proportionality constant depends on the value of the parameters k and c.

2. The general exponential function can also be written in the form $y = k \cdot b^x$, where k is some nonzero number and b is a positive number other than 1. (What happens if b is equal to 1?) In this form, too, the function has the proportionality property.

   Investigate how the value of the proportionality constant depends on the value of the parameters k and b.

Interior Photography

3 Los Altos High School, Judy Strauss, Lynne Alper; **4** PhotoDisc; **5** Leo de Wys Inc. PHOTOFILE; **10** Kapa'a High School, Elaine Denny; **14** FPG International; **20** Santa Cruz High School, Kevin Drinkard, Lynne Alper, **21** Comstock ©1994; **31** Leo de Wys Inc. PHOTOFILE; **34** Lincoln High School, Lori Green; **36** Leo de Wys Inc. PHOTOFILE; **38** PhotoEdit; **41** Stock Boston; **49** Brookline High School, Terry Nowak; **50** Hillary Turner; **51** Hillary Turner; **52** SuperStock, Inc.; **57** PhotoEdit; **63** SuperStock, Inc.; **65** Comstock ©1989; **66** Santa Cruz High School, Kevin Drinkard, Lynne Alper, **67** FPG International; **69** Stock Boston; **70** SuperStock, Inc., **78** San Lorenzo Valley High School, Dennis Cavaillé, Lynne Alper; **81** PhotoDisc; **83** Hillary Turner; **89** The Image Bank; **100** SuperStock, Inc.; **105** SuperStock, Inc.; **106** SuperStock, Inc.; **109** SuperStock, Inc.

Cover Photography

Fireworks Image Bank; *Pennant Fever* SuperStock, Inc.; *Meadows or Malls?* SuperStock, Inc.; *Small World, Isn't It?* Hillary Turner and Richard Wheeler; *Orchard Hideout* Hillary Turner and Richard Wheeler

Front Cover Students

Small World Isn't It?, first row: Allan Duncan, Laura Holst, Alycia Brown, Neema Patel, Nick Bulgerin, Arthur Alcones, Efigenia Vasquez. Second Row: Rey Swengel, Amie Yeargan, Simone Adams, Issac Dozer, Ali Pourghadir, Kong Phan, LaQuana Lee, Charmayne Alcones. *Orchard Hideout,* from left, Sung Lee, Jordan Bromely, Trenika Fields.